7th International Conference on Local Knowledge (ICLK 2017)

Engaging Local Wisdom in the New Millenium

SHS Web of Conferences Volume 45 (2018)

Johor Bahru, Malaysia
8-9 May 2017

Editors:

Nurul Farhana Low Abdullah
Norizan Esa

ISBN: 978-1-5108-6468-9

Printed from e-media with permission by:

Curran Associates, Inc.
57 Morehouse Lane
Red Hook, NY 12571

Some format issues inherent in the e-media version may also appear in this print version.

Printed by Curran Associates, Inc. (2018)

For additional information, please contact EDP Sciences – Web of Conferences at the address below.

EDP Sciences – Web of Conferences
17, Avenue du Hoggar
Parc d'Activité de Courtabœuf
BP 112
F-91944 Les Ulis Cedex A
France

Phone: +33 (0) 1 69 18 75 75
Fax: +33 (0) 1 69 28 84 91

contact-edps@webofconferences.org

Additional copies of this publication are available from:

Curran Associates, Inc.
57 Morehouse Lane
Red Hook, NY 12571 USA
Phone: 845-758-0400
Fax: 845-758-2633
Email: curran@proceedings.com
Web: www.proceedings.com

TABLE OF CONTENTS

SHS Web of Conferences **45**, 00001 (2018) https://doi.org/10.1051/shsconf/20184500001
ICLK 2017

Preface – Engaging Local Wisdom in the New Millenium

Nurul Farhana Low Abdullah[1] [1]
[1]School of Humanities, Universiti Sains Malaysia

The United Nations Educational, Scientific and Cultural Organisation (UNESCO) describes Local knowledge as 'the understandings, skills and philosophies developed by societies with long histories of interaction with their natural surroundings'. In rural and indigenous communities, local knowledge 'informs decision making about fundamental aspects of day to day life' and forms the basis for the various facets of culture in the community including language, social interactions, ritual and spirituality, as well as practices related to use of resources. [1]

This body of wisdom has developed over time and continues to develop. The practical aspects of local knowledge are highlighted by the UN Food and Agriculture Organisation through its definition of local knowledge: '…the entire system of concepts, beliefs and perceptions that people hold about the world around them. This includes the way people observe and measure their surroundings, how they solve problems and validate new information. It includes the processes whereby knowledge is generated, stored, applied and transmitted to others'. [2]

The fact of these formulations being endorsed by a global organisation like the UN validates and necessitates further study into Local Knowledge, its universal relevance and multi-faceted applications across disciplines and locations. Thus, in the spirit of inclusiveness and inquiry. the Local Knowledge Study Group of Universiti Sains Malaysia (USM) first embarked on its quest to foreground the study of this transdisciplinary form of knowledge at regional level to serve as 'a foundation for locally-appropriate sustainable development' (as succinctly noted by UNESCO) with its inaugural Regional Conference on Local Knowledge in Langkawi, Kedah in 2011. Seven annual conferences and 40 edited books later, the organisers of the (rebranded) International Conference on Local Knowledge (ICLK) 2017 are seeking to disseminate contemporary research findings, built around the theme of "Engaging Local Wisdom in the New Millenium" to a global audience through its first publication of indexed proceedings in SHS Web of Conferences.

The seventh instalment of ICLK was organised by USM on 8-9 May 2017 in collaboration with the Johor State Government through its cultural arm, the *Yayasan Warisan Johor* and supported by *Dewan Bahasa dan Pustaka* where 90 papers were presented by local and international participants. The papers in these Proceedings represent a selection of the latest transdisciplinary research that engages with various forms of Local Wisdom in their framework of analysis. The articles have been divided into the following areas, according to thematic focus: Architecture and Conservation, Arts and Crafts, Ecology and Ecosystems, Governance and Policy, Health and Healing, History and Community and finally, Language, Literature and Media.

As guest editor of this volume of selected conference papers, I would like to take this opportunity to thank the members of the Scientific Committee, reviewers and contributors for their co-operation and commitment throughout the process of producing this collection of papers. I would also like to record sincere thanks to all parties who participated in the organisation of the 7th ICLK for making the conference a success.

GUEST EDITOR

Nurul Farhana Low Abdullah, *Universiti Sains Malaysia*

[1] Corresponding author: nflow@usm.my

REFERENCES

1. UNESCO – Natural Sciences – Priority Areas: http://www.unesco.org/new/en/natural-sciences/priority-areas/links/related-information/what-is-local-and-indigenous-knowledge/

2. Food and Agriculture Organization of the United Nations – Fact Sheet: http://www.fao.org/docrep/007/y5610e/y5610e01.htm#bm1

Statement of Peer review

In submitting conference proceedings to *Web of Conferences*, the editors of the proceedings certify to the Publisher that

1. They adhere to its **Policy on Publishing Integrity** in order to safeguard good scientific practice in publishing.
2. All articles have been subjected to peer review administered by the proceedings editors.
3. Reviews have been conducted by expert referees, who have been requested to provide unbiased and constructive comments aimed, whenever possible, at improving the work.
4. Proceedings editors have taken all reasonable steps to ensure the quality of the materials they publish and their decision to accept or reject a paper for publication has been based only on the merits of the work and the relevance to the journal.

Title, date and place of the conference

7th International Conference on Local Knowledge (ICLK 2017) May 8-9, 2017, Johor Bahru, MALAYSIA

Proceedings editor(s):

Nurul Farhana Low binti Abdullah

Norizan binti Esa

Date and editor's signature

21 May, 2018

SHS Web of Conferences **45**, 01001 (2018) https://doi.org/10.1051/shsconf/20184501001
ICLK 2017

Thermal Surface Analysis on Neo-minimalist apartment façades in Penang, Malaysia

Yasser Arab[1,1], *Ahmad* Sanusi Hassan[1], and *Bushra* Qanaa[2]

[1]School of Housing, Building, and Planning. Universiti Sains Malaysia
[2]Faculty of Architecture, Ittihad Private University

Abstract. The study investigates apartment's façade thermal performance with neo-minimalist architectural style in Penang, Malaysia. Neo-minimalist style is considered as the most popular style in Malaysia in 2010s. The style is rediscovering from early modern minimalist movement with a design concept "less is more". It applies minimal and efficient design of architectural character in defining form and space. Penang Island the second most important city in Malaysia after Kuala Lumpur. It is located at the north-western part of the country. The first case studies is the Light Linear apartment which has sixteen stories located on the east cost of Penang Island at Pantai Street, Penang. The second case study is BayStar apartment building, the eleven stories building tis located in Bayan Lepas at the seaside facing Jerejak Island. In order to conduct this study Fluke Ti20 thermal imager was used to capture thermal images for the west facades of the selected case study hourly from 12:00 to 6:00 pm on 15th March 2017. The study finds that the recessed wall, balconies and the shading devices were the important elements to provide shades on the façades for good thermal performance.

1 INTRODUCTION

The study aims to investigate the façade surface temperature of two neo-minimalist architecture style high-rise buildings within Penang Island, Malaysia. The two selected case studies are located in the eastern part of Penang Island, Malaysia with longitude 100.3288° E and latitude 5.4142° N. Penang is considered the second most important state in Malaysia [1, 2], and it has experienced rapid development in the last three decades [3]. High-rise buildings have been built in Malaysia since the 1970s. Lately, high-rise buildings have become a popular residential building type in Malaysia because of the ability to provide the required high number of dwelling units [4]. A 2010 study by the Department of Statistics Malaysia found that over 2 million people live in high-rise apartment buildings in Malaysia, and this number has been increasing rapidly in the last 2 decades [5] (Figure 1).

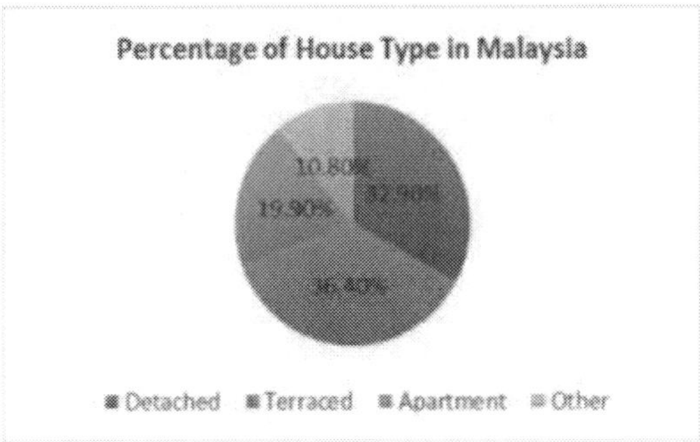

Fig.1. Houses types ratio in Malaysia.

[1] Corresponding author: yasserarab2005@yahoo.com

2 PROBLEM STATEMENT

Energy consumption is considered one of the most important issues that scholars are trying to solve and reduce. In countries with tropical climate such as Malaysia, it is very important to have an excellent façade design to be able to provide sufficient shading performance and prevent unnecessary solar radiations from penetrating into the house [6–8]. When the building façade is exposed to direct sunlight which is the case for many buildings, it has a problem of transferring the heat from outside wall to the indoor area of the building. The high amount of unnecessary solar radiation causes heat gain inside the building which leads to uncomfortable thermal conditions due to the temperature increase which growths the usage of air-conditioning to cool down indoor area [9–11].

3 CASE STUDIES

The first selected case study is The Light Linear. The sixteen stories building is located on the east coast of Penang Island at Pantai Street, Penang (Figure 2). While the second selected case study is BayStar apartment which is eleven stories tall and is located by the seas, facing Jerejak Island in Bayan Lepas. Both selected case studies have neo-minimalist architectural style, and they are built for high-income families (Figure 3)

Fig. 2 The light Linear Condominium

Fig. 3 Baystar Condominium

4 METHODOLOGY

This research is based on field study. It involves snapping thermal images of the west façade for the two selected case studies. The thermal images were taken by thermal camera known as Fluke Ti20 thermal imager (Figure 4). The thermal images were snapped hourly base from 12:00 to 6:00 pm on 15th March 2017 [12].

The Fluke Ti20 is an infrared camera which is able to take images with thermal information that can provide the temperature of every point of the building façade [13]. To analyse the thermal images taken by the Fluke Ti20, the images were then exported into SmartView® computer software which comes with the Fluke device. This software has tools to help the researcher to analyse the images and the results, and produces photos with infrared and digital images together to provide all the required information in a very easy way to perform the analysis [14, 15] (Figure 5).

Fig. 4. Fluke Ti20 thermal imager

Fig. 5. An example of an infrared image

5 RESULTS ANALYSES

Three limitations were involved in this study, the first limitation is that the study was conducted in the afternoon hours as mentioned earlier (from 12:00 to 6:00 pm), and this was because the sunlight during the morning hours is not as intense as during the afternoon hours, and is more important to be measured. The second limitation of the research was a delay time when taking the thermal images of the two selected case studies as the two buildings are located about 5 kilometres apart from each other. It caused about 10 minutes delay. While the other limitation is the two selected case studies are of different heights, as they are 11 and 16 stories tall respectively.

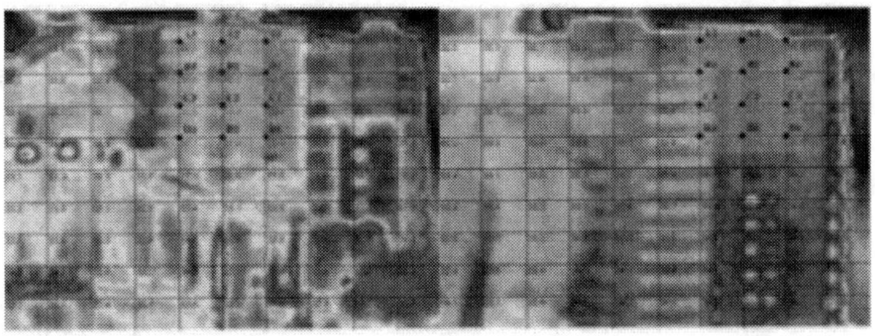

Fig. 6. The selected points of both case studies. (Left: Baystar. Right: The Light Linear)

As mentioned earlier, the thermal images were taken hourly for the two selected case studies from 12:00 to 6:00 pm on 15th March. Figure 6 shows the facade selected points of both case studies. The resulting surface temperature for the selected points are shown in Table 1 -3 and Figure 7 as follows:

Table 1. The thermal images of the two selected case studies on 15th March 2015.

Time	**Baystar**	**The Light Linear**
12:00		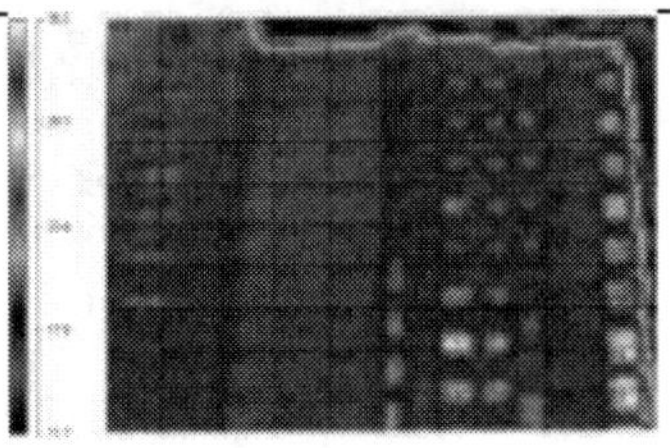

1:00	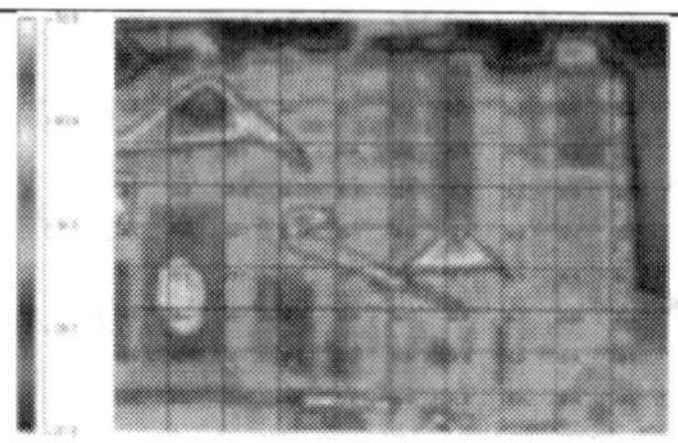	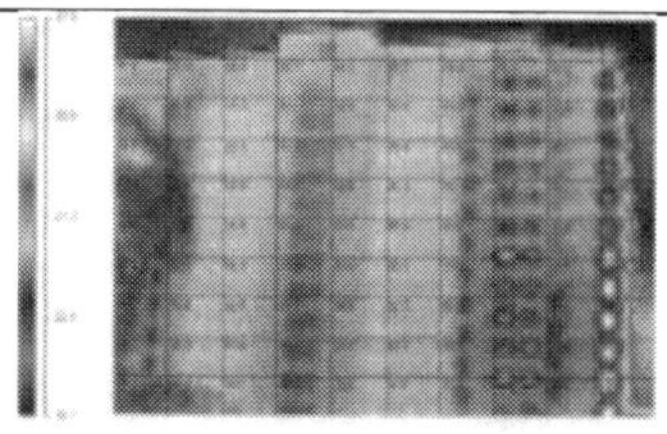
2:00		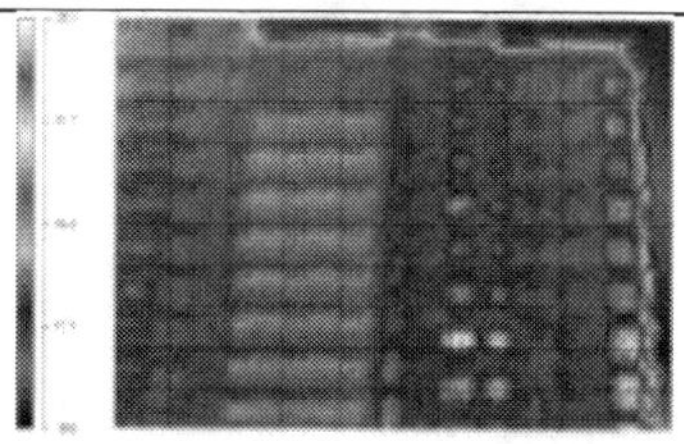
3:00		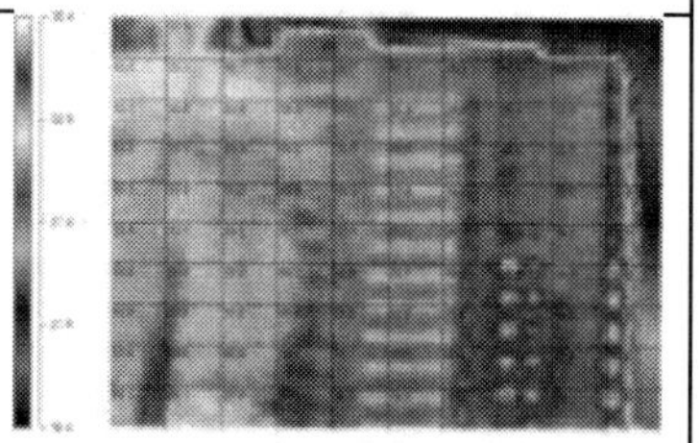
4:00		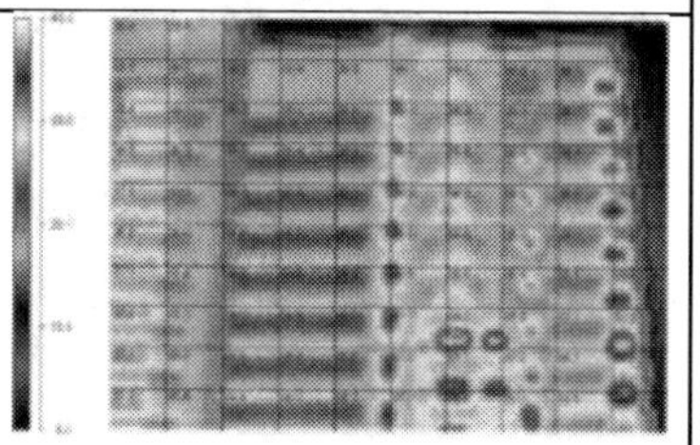
5:00		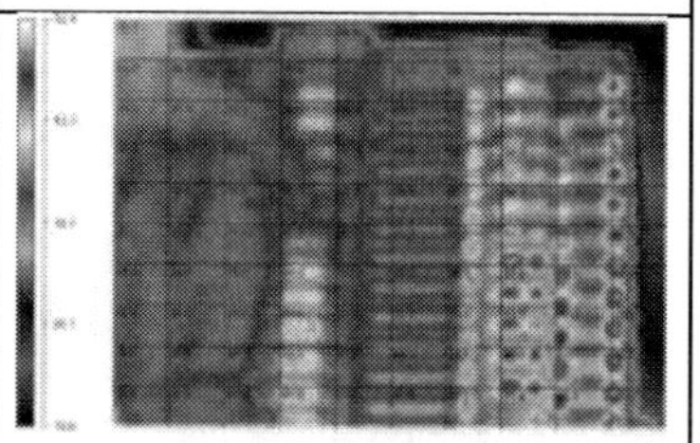
6:00		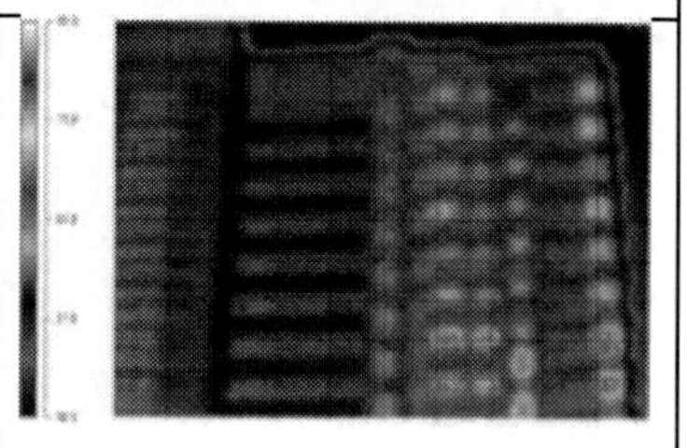

Table 2. The surface temperature of Mutiara Idaman 1 west façade selected points.

		Point / Time	A1	A2	A3	B1	B2	B3	C1	C2	C3	D1	D2	D3
Baystar	15-Mar	12:00	32.2	32.3	32.8	32.7	32.3	32.8	32.6	32.4	33	32.7	32.4	33.1
		1:00	32.3	32.5	33.2	33	32.7	33.1	32.8	32.9	33.4	32.8	32.8	33.7
		2:00	33.5	33.2	34.1	34.6	33.5	34.1	34.3	34	34.4	34.6	34.2	35.2
		3:00	33.4	33.6	34.5	35.4	33.9	34.4	35.3	34.2	34.6	35.3	34.4	35.2
		4:00	36.6	35.6	36.7	37.8	35.8	37	37.6	35.9	37.4	38.3	36.7	37.8
		5:00	37	36.5	38.1	39	36.5	37.7	38.7	36.7	38.5	38.8	37.3	39.4
		6:00	35.6	35.8	37.9	36.6	36	37.8	36.9	36.6	38.9	36.9	37.1	39.6

Table 3. The surface temperature of The Light Linear west façade selected points.

		Point / Time	A1	A2	A3	B1	B2	B3	C1	C2	C3	D1	D2	D3
The light Linear	15-Mar	12:00	31.3	31.6	31.6	31.4	31.9	31.7	31.4	31.8	31.7	31.4	31.8	31.8
		1:00	32.9	33.3	32.9	32.9	33.3	33	33.1	33.6	33	33.1	33.3	33.1
		2:00	34.1	33.9	34.7	34.1	34.1	34.5	34	34.1	34.5	34.2	34.2	34.8
		3:00	34	34.5	34.2	33.9	34.2	33.9	34	34.4	34	34.2	34.3	34.1
		4:00	36.1	34.9	36.2	36.1	34.9	35.8	35.6	35.6	35.9	36	35.9	36.3
		5:00	38.2	39.3	38.7	38.6	39	38.5	39.3	40	38.7	39.3	39.5	38.9
		6:00	40	38.2	40.8	40.1	39	40.8	39.4	39.4	40.7	39.6	39.5	41.4

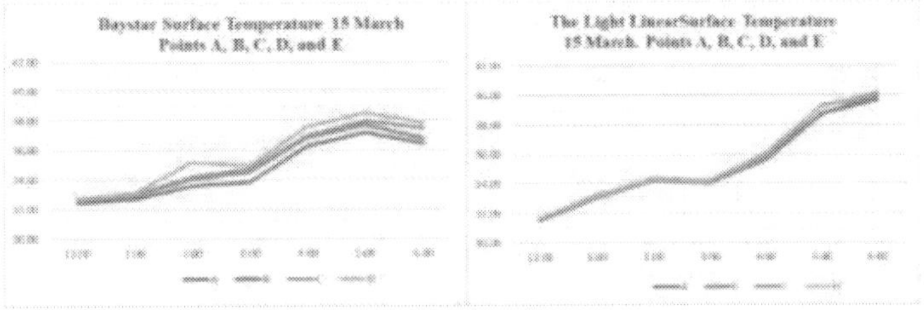

Fig. 7. The façade selected points' surface temperature of both case studies (Left: Baystar. Right: The Light Linear).

6 DISCUSSION

Figure 8 shows the average of façade surface temperature for both selected case studies, where the findings show that:

- Both Baystar and The Light Linear have similar temperature readings in most of the measurement hours.
- The highest temperature was at 6:00 pm at The Light Linear with 39.91°, while the lowest temperature was 31.62° during the first hour.
- The lowest temperature measured at Baystar façade was 32.61° at 12:00 pm and the highest temperature was 37.85° at 5:00 pm.
- Comparing the results with average façade surface temperature of modern architectural style buildings (figure 9), the results show that the neo-minimalist style buildings have better shading performance.

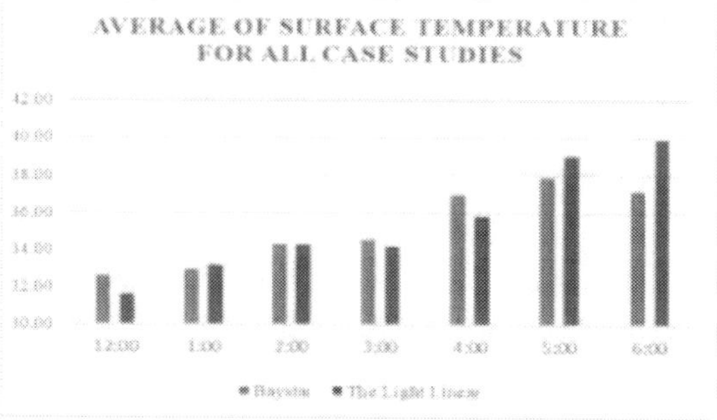

Fig. 8. The Average surface temperature of both case studies

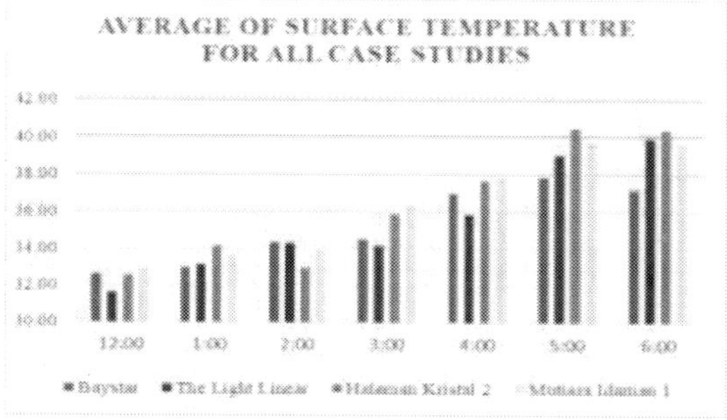

Fig. 9. The Average surface temperature of neo-minimalist and modern style buildings.

7 CONCLUSION

The research outcomes show that neo-minimalist architectural style is efficient, and it is able to provide good façade shading elements like wide balconies, recessed wall and preventing the unnecessary intense sunlight and solar radiation from increasing the indoor atmosphere temperature, and that will help to achieve the indoor thermal comfort. So, applying this architectural style will help to decrease the use of mechanical air conditioning for the thermal comfort. This research is the first step in the guide for designers and architects for their upcoming projects in order to design a building with efficient façade shading design.

8 ACKNOWLEDGMENT

The authors would like to express appreciation for the financial support under Research University Grant by Universiti Sains Malaysia.

9 REFERENCES

1. Y. Arab. "Facade Design Efficiency on Extent Sunlight Penetration in Neo-Minimalist Style Apartments in Penang, Malaysia", Journal of Architectural Engineering Technology, Vol. **4**(2) (2015).
2. A. S. Hassan. Y. Arab, and Mazran Ismail. "Architectural Styles and Developments of Apartments in Putrajaya, Malaysia." (2015).
3. Tan, K. C., San Lim, H., MatJafri, M. Z., & Abdullah, K. Landsat data to evaluate urban expansion and determine land use/land cover changes in Penang Island, Malaysia. Environmental Earth Sciences, **60**(7), 1509-1521, 2010
4. Hoffman, A. v. High ambitions: The past and future of American low-income housing policy. Housing Policy Debate, **7**(3), 423-446, (1996).

5. Malaysia, D. o. (2010). Characteristics of Living Quarters 2010. Putrajaya: Department of Statistics Malaysia.
6. A. S. Hassan and M. Ramli, "Natural ventilation of indoor air temperature: A case study of the Traditional Malay House in Penang", American Journal of Engineering and Applied Sciences, pp. 521-528, (2010).
7. H. Feriadi and N. H. Wong, "Thermal comfort for naturally ventilated houses in Indonesia", Energy and Buildings, pp. 614-626, (2004).
8. M. S. Bakhlah and A. S. Hassan, "The study of air temperature when the sun path direction to ka'abah: with a case study of Al-Malik Khalid Mosque, Malaysia", International Transaction Journal of Engineering, Management & Applied Sciences & Technologies, pp. 185-202, (2012).
9. Rahman, A. M. Housing design in relation to environmental comfort: A comparison of the traditional Malay house and modern housing including work in the tripartite programme between Universiti Sains Malaysia, the Welsh School of Architecture and the UK building research. Building research and information, **23**(1), 49-54, (1995).
10. Royle, K., & Terry, C. Hawaiian Design: Strategies for Energy Efficient Architecture. Collingdale, Honolulus: DIANE Publishing, (1990).
11. K. M. Al-Obaidi, M. Ismail and A. M. A. Rahman, "A review of the potential of attic ventilation by passive and active turbine ventilators in tropical Malaysia", Sustainable Cities and Society, Vol.**10**, pp. 232-240, (2014).
12. A. S. Hassan, and Y. Arab. "Thermal Façade Surface Study on Early Modern Apartments in Kuala Lumpur". Journal of Built Environment, Technology and Engineering, Vol. **2** (2017).
13. Bezbabicheva, O. I., Bilchenko, A. V. and Kyslov, A. H. "Forecasting of Temperature Tension in Facade Beams of Bridge Constructions". Science and Transport Progress. Bulletin of Dnipropetrovsk National University of Railway Transport, **33**, PP 28-31, (2010).
14. Badida, M., Dzuro, T., & Franko, Š. Utilization infrared thermography in the automotive industry to determine faults by using thermal camera Fluke Ti10. Глобальне управління та економіка, **1**, 7-13, (2015).
15. Županić, D., Bolarić, A., Stančić, M., Piljek, D., & Jurinjak, B. (2015). Thermal imaging of the outer surface of the Laboratory for engines and vehicles at the Faculty of mechanical engineering and naval architecture. Interklima (2015).

The Traditional Architecture of the Melanau Tall Longhouse, Mukah, Sarawak

Azizi Bahauddin [1,] and *Dyg Mustika Syaheeda* Awg Musadat [2]

[1, 2] School of Housing, Building and Planning, Universiti Sains Malaysia, Penang, Malaysia

Abstract. Full of history and legends, the Melanau, one of the ethnic groups of Mukah, Sarawak, Borneo, went through a few periods of cultural assimilation. Three major areas: the Tellian Daya (Ulu), Tellian Puloh and Tellian Laut communities, are the main enclaves of these people with Tellian Puloh being the main hub. Their settlements along the river banks are in lowlands within the tidal zone. During the monsoon period (*lian suluh*), the tides combined with the flood water from the hills produce unpredictable fluctuations of high and low water and flooding creating a saturated bog like quality. This paper aims to investigate the Melanau tall longhouse. Its construction was built in such tall heights due to the conditions of the soft soil. Often known to have been built up to 40 feet high, the house also provides protection for the people from floods as well as what used to be their enemies of other warring tribes. The materials used for the construction of the house came from the surrounding areas of sago leaves for the roof, sago tree trunk that had been sliced into planks for walls, *nibong* tree trunks for the floor and *belian* (Bornean iron wood) tree trunks for the main structural components of the house. All of these materials are known to be hardy and water resistance. Unfortunately, these traditional houses no longer exist with a few being abandoned and left to rot. Architectural documentation of this house is urgently needed to record the historical part of this culture. The investigation takes on qualitative method looking into the cultural influence onto the architecture. Survey, observation, architectural documentation and interviews with the Melanau cultural experts were undertaken. The finding pointed towards the establishment of the architectural language of the house based on the culture influence and functions of the spaces.

1 Introduction

Sarawak (Fig. 1) is the largest state in Malaysia (Fig. 2) and known by its various ethnic groups. Among ethnic groups in Sarawak are Malay, Chinese, Indian, Iban, Bidayuh and Melanau. The Melanau is viewed among the earliest settlers in Sarawak and is the fourth Sarawak predominant ethnic, after Malay, Iban and Bidayuh. The Melanau live alongside the lower reaches of the Rajang River, the well-known river in Sarawak, and spread alongside the coast from the mouth of the Rajang northeast, to Bintulu, and from the coast of Kanowit. Due to that, they call themselves *a-likow*, meaning 'people of the river'.

Fig.1. Map of Sarawak

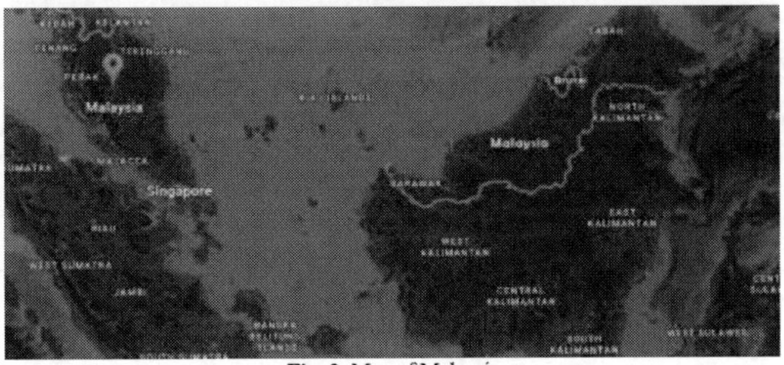

Fig. 2. Map of Malaysia

In ancient times, the Melanau used to stay in tall and prolonged houses, but nowadays, they live in Malaysian kampung-style properties. The Melanau traditional house often known to be 40 feet high from the ground and built by using solid *belian, nibong* and sago tree trunks. In the past, a hamlet of Melanau usually consist of one to three tall longhouses, located in strategic position; at the river mouth within the tidal zone by taking into consideration the security and economic activities. The Melanau community worked as boat builders. However, their main jobs are as *padi* farmers, fisherman and most importantly, as sago farmers [1]. The sago trees commonly found in freshwater swamps around Mukah, Oya, and Dalat. Since the Melanau have constantly been the most distinguished inhabitants of the southwest coastal area, in order to manipulate the revenue from the sago trade, as early as 400 years in the past, the Sultanate of Brunei appointed Melanau representatives at the mouths of the most necessary sago-producing rivers, especially Oya River and Mukah River [2].

2 Methodology

The research utilized a qualitative method relying on ethnography and phenomenology documentation upheld by a top to bottom examination in the Tellian Puloh, Mukah. It will be supported by looking into the cultural influence onto the architectural language of the Melanau Tall Longhouse. Data obtained through observation, architectural documentation on the longhouse and interviews with the Melanau cultural experts. Architectural documentation of this house is urgently needed to record the historical part of this culture. Therefore, background research on few recognizes longhouses conducted before further work can occur. Nonetheless, the absence of optional information requires the analyst to rely on primary sources. This research included few non-structured interviews with few longhouses owner at Tellian Puloh. This procedure ran at the same time with the visual information accumulation on their traditional house and ways of living. For this paper, only a single case study explored which is the Melanau Tall Longhouse in Sarawak Cultural Village. This process will capture the architectural language of an actual Melanau Tall Longhouse.

3 The Melanau Community

Melanau ethnic is believed to have originated from the island of Java as their traditional clothes seem to have Java influences. From the aspects of the spoken language and the practice of pre-Islam, Melanau ethnic shows similarity with the Orang Ulu ethnic; Sekapan, Lahanan, Punan Ba, Kanowit from Belaga in Hulu Rejang, Kanowit and Tanjung in Central Rejang. Melanau have their own spoken language with vary dialect and they can be differentiated among themselves through their dialect. Although each community has its own attribute dialect, they all share the equal cultural and lingual background. However, there is a small percentage of this community speak local Malay dialect. This group of Melanau is all Muslims. They embrace the Malay's way of life, yet some of the Melanau cultures are still being practiced. Melanau dialect was usually used as a part of discussion and their traditional use freely either in entertainment or singing.

Years ago, the Melanau practiced animist religion before Islam and Christian religions came to Sarawak. Islam religion spread throughout the Melanau region; Mukah, Oya, Dalat, Matu, Rejang, Sibu, Igan, Sarikei, Tatau, Bintulu and Balingian; by the Sultanate of Brunei. The Melanau embraced Islam because they were captivated by the teachings of Islam and it was also a part of the basis ruling, to respect the Sultanate [3]. With regard to Christianity, the religion spread through the expansion of British rule in Sarawak. The British built churches at each region, Catholic schools monitored by Fathers, Mothers, and Sisters, along with introducing the tax system [4].

Whilst the Melanau was originally animist follower, a majority of the Melanau nowadays are Muslim, although some of them, particularly among the Melanau Mukah, and Dalat are Christian. In spite of that, most of the Melanau still practice their ancestor traditional rites such as the Kaul Festival. In spite of their different beliefs and religions, the Melanau are very tolerant among themselves. One can still come across a Melanau family with different family members embracing different religions whilst their parents nonetheless have robust animist beliefs.

4 Case Study- Melanau Tall Longhouse, Sarawak Cultural Village

Due to the fact that these traditional houses no longer exist within Tellin Puloh perimeter, with a few being abandoned, a Melanau Tall Longhouse model in Sarawak Cultural Village (Fig. 3) is chosen as the main site for this case study. Even though it was built for tourism purposes and is located in the Kuching City, through first observation, the overall architectural language of the model can be considered authentic.

Fig. 3. Melanau Tall Longhouse, Sarawak Cultural Village

The Melanau traditional house is a type of a longhouse but it is famously known by its' height. Therefore, it is named as the tall longhouse and consists of three levels; first level, main level, and upper level. The Melanau Tall Longhouse model in Sarawak Cultural Village is a living museum that is located at the foothills of Mount Santubong, about 35 km from Kuching. Besides functioning as a living museum, the management provides an area that function as a homestay for tourists all around the world. Every area of the house is loaded with tradition artifact belonged to the Melanau tribe. The structures of this house built using conventional materials which are readily available from the jungle and all of these materials are known to be hardy and water resistance. *Belian* tree trunks used as the main structural components of this house especially for the columns (Fig. 4) and the stairs. The wall structure and flooring are made out of *nibong* tree trunks. Lastly, the main economical source for the Melanau community, sago leaves, is use to build the roof, while the trunks will be sliced into planks and can be used as a part of the wall materials.

Fig. 4. Solid *Belian* Tree Trunks as Columns

Melanau Tall Longhouse has its own historical background, contradict to other traditional longhouses in Sarawak. The building structure and height can be named as part of the uniqueness of this building. Unlike other traditional longhouses in Sarawak, Melanau Tall Longhouse (Fig. 5 and 6) is built 40 feet high from the ground and consists of 20 to 30 living spaces for each family staying at that tall longhouse [5].

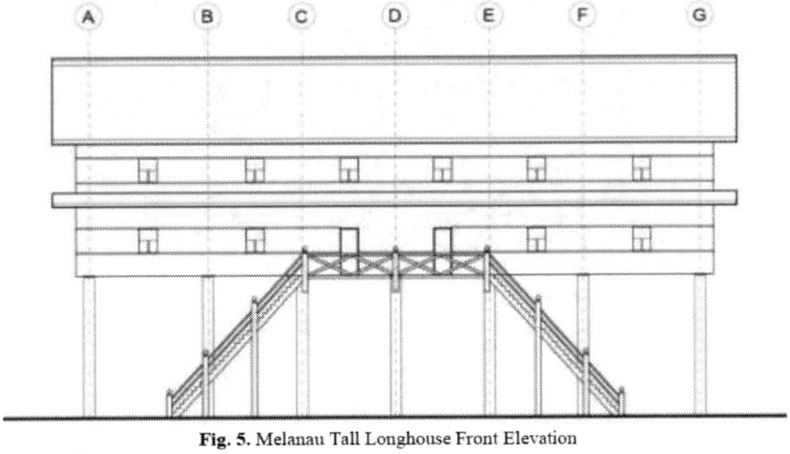

Fig. 5. Melanau Tall Longhouse Front Elevation

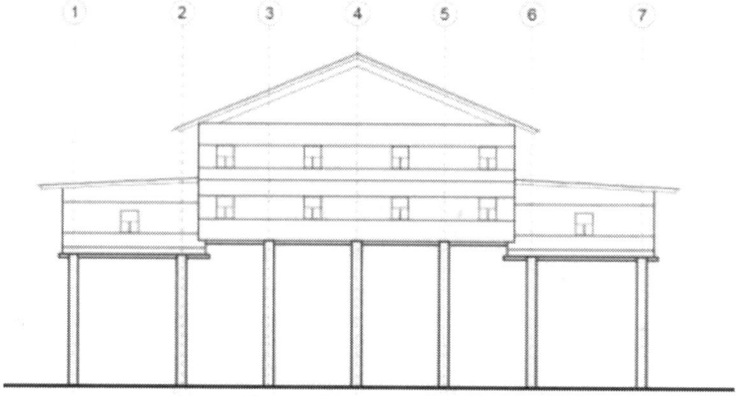

Fig. 6. Melanau Tall Longhouse Rear Elevation

Melanau traditional house flooring is made out of double layer *nibong* (Fig. 7) trunks and are arranged in a criss-cross pattern with tiny gaps in between, making sharp weapons such as lances and blades to hardly pass through. Those tiny gaps work as a part of the house protection from their enemies of other warring tribes, especially from the Headhunters, the Iban ethnics. The Melanau community will pour boiling water onto the floor when attacked by other tribes from below.

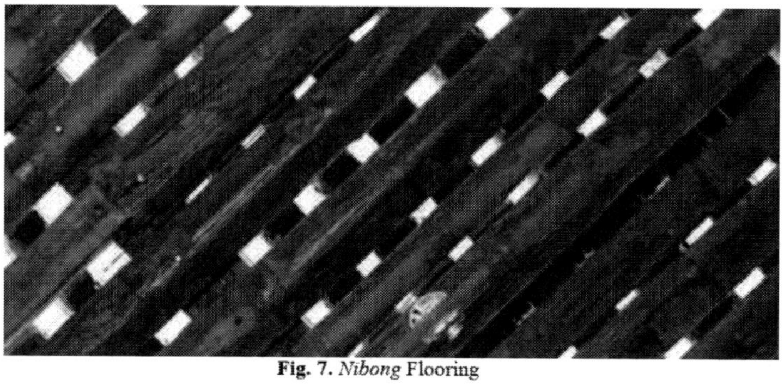

Fig. 7. *Nibong* Flooring

Fig. 8. Main Level Floor Plan

Fig. 9. Upper Level Floor Plan

The main floor (Fig. 8) of this house act as a mutual space for every family and is also where the kitchen is located. The centre space of this building function as the family area, where any types of ceremony from activity of life starting until the person is dead took place. Main floor consists of few rooms, 20 to 30 apartments, for every family that are staying in that tall longhouse. Next, the upper level (Fig. 9) is where the Melanau keep their tradition artifact and family's assets.

5 Conclusion

This paper examines and analyses the longhouse associated with the Melanau Tall Longhouse as the focus of the study in relation to their traditional culture. The research looks into the understanding of their unique longhouse which is differ to other longhouses in Sarawak. The complex nature of the Melanau community established the concepts of sharing and cooperation within the overall layout design of the Melanau Tall Longhouse. It is cultural conventions and cues that control the information linking social organisation with the built environment with the adoption of local materials as the main structure of the building. Regardless of its extinction, the interpretation received from the overall view on its architectural language has demonstrated the meaning of this longhouse from top to bottom and the importance of having a sheltered for this ethnic family. The Melanau convention culture offers boundless legacy values that are valuable for the general population and also guests to comprehend this unique group.

6 Acknowledgement

The authors would like to thanks Ministry of Higher Education of Malaysia for providing Fundamental Research Grant Scheme (FRGS - 203 / PPBGN / 6711554) to support this study. The credit also goes to the School of Housing, Building and Planning, Universiti Sains Malaysia for providing facilities to carry out the work.

References

1. S. Baring-Gould, and C.A. Bampfylde, A history of Sarawak: *under its two white rajahs, 1839-1908*, **20** (2007)
2. S. Morris, The Oya Melanau., (1978)
3. P.L. Amin Sweeney, Journal of the Malaysian Branch of the Royal Asiatic Society, *Silsilah Raja-raja Brunei*, **1-82** (1968)
4. Dzulfawati Hassan, Upacara Kematian Melanau Likow di Sarawak, (2006)
5. V.F. Chen, The Encyclopaedia of Malaysia: *Architecture*, (1998)

The Importance of Scientific Tests In Masonry Heritage Building Conservation In Malaysia

Tan *Chin Ling*[1,*], and *A Ghafar* Ahmad[1]

[1]School of Housing, Building and Planning, Universiti Sains Malaysia, 11800 Penang, Malaysia

Abstract. Since the Portuguese era in Malacca in the 16th century, many authentic colonial heritage buildings were built in Malacca and other parts of Malaysia, most of which were masonry structural buildings. After Melaka and George Town received recognition from UNESCO as World Heritage Sites on 7th July 2008, heritage building conservation in Malaysia has attracted serious attention from various parties. Since the establishment of the National Heritage Department under the National Heritage Act 2005 (Act 645) on 1 March 2006, the implementation of scientific tests in heritage building conservation practice in Malaysia has been encouraged and recommended by the Department. However, scientific tests are normally only conducted in conservation works of National Heritage buildings and some Category I heritage buildings, but not made compulsory for all heritage buildings conservation projects in Malaysia. This paper is based on a research carried out to study the scientific tests conducted in masonry heritage building as part of conservation practice in Malaysia. The research findings revealed that scientific tests such as X-ray fluorescence test, ion chromatography test, moisture test and Schmidt hammer rebound test are important for masonry heritage building conservation in Malaysia. The basic guidelines to conduct the scientific tests derived from the research findings will be able to enhance the current practice in conserving masonry heritage buildings in Malaysia.

1 Introduction

Several studies related to scientific tests such as X-ray fluorescence test, ion chromatography test, moisture & dampness test and Schmidt hammer rebound test in heritage building conservation have been carried out in Europe, Australia, Malaysia and other countries. A Ghafar and Haris Fadzilah. (2010) revealed that salt attack and rising dampness are the common building defects in heritage buildings in Penang, Malaysia. The paper stated that the existence of mineral salts may weaken the building components. The three types of destructive soluble salts which commonly found in the masonry wall of heritage buildings were chlorides (Cl), nitrates (NO_3) and sulphates (SO_4). These soluble salts can be detected by conducting ion chromatography test.[1] Paulo Lourenco et al.(2014) revealed the possible causes of damage and the usage of scientific expert systems in building defects diagnosis by understanding the chemical, physical and properties of mortar, brick and masonry.[2] Nadia Bianco et al. (2013) reported the evaluation of the performance of the mortar by tracing on the basis of the results of the analyses and tests. In the other words, scientific test is an effective tool to diagnosis building defect and building material performance.[3] A Ghafar Ahmad et al. (2005) stated that several scientific tests such as x-ray fluorescence test, salt level test, Schmidt hammer rebound test and paint colour scheme analysis were carried out in Old City Hall, George Town building conservation project to safeguard its authenticity and historical value.[4] These papers showed that scientific tests play an important role in heritage building conservation. In the Technical Guide of Salt Attack and Rising Damp, a guide to salt damp in historic and older buildings, David Young (2008) found that accurate diagnosis is critically important where there is no substitute for a thorough understanding of a building's behavior and its response to changes over time. It only can be done by conducting systematic scientific tests.[5]

2 Methodology

The masonry heritage building which is the major typology of heritage building in Malaysia has been selected as research building type for this study. The data was collected from several literature reviews and documents such as final conservation reports and scientific test reports conducted in masonry heritage building conservation projects in Malaysia. The data was compiled and analysed to reveal the importance of the scientific test in masonry heritage building conservation. Furthermore, the basic guidelines to conduct the scientific tests were derived from the research findings with the aim to enhance current practices.

3 Discussion and Findings

3.1 Significance of Risk in Building Diagnostics

Ana Pereira R. and Douglas J. (2008) stated that all interventions in the built environment incur risks.[6] The word 'risk' has been defined as the combination of possibility of an event and its consequence.[7] Hence, in heritage building conservation, application of appropriate conservation methods or techniques is the key towards ensuring a successful conservation project. Therefore, the decision making tool in heritage building conservation projects should be evidence-based practice involving the application of scientific tests. Evidence is something that is visible, measurable and recordable.[8] Evidence is the facts, information and data that answers the question regarding building failures. The evidence can be observed directly on site or revealed by conducting scientific tests. The quantity or quality of evidence is related to the significance of risk in building diagnostics. For instance, no information may lead to uncertainty, while comprehensive information may lead to certainty.[9] According to Douglas J. (2013), "*risk always involves an element of uncertainty, but uncertainty does not always involve risk.*"[10]

3.2 Scientific Methods

According to the Oxford Dictionaries Online, scientific method is defined as "*A method of procedure that has characterized natural science since the 17th century, consisting in systematic observation, measurement, and experiment, and the formulation, testing, and modification of hypotheses.*"[11] In heritage building conservation, building conservators or architects may require the expertise of specialists to detect or analyse the building defects. These specialists such as chemists or material engineers will conduct scientific methods to assist in building diagnostics. There are several scientific methods or tests to assist building diagnostics, including destructive and non-destructive tests.

3.3 X-ray Fluorescence (XRF) Test

X-ray fluorescence (XRF) test refers to "*the emission of characteristic "secondary" (or fluorescent) X-rays from a material that has been excited by bombarding with high-energy x-rays or gamma rays.*"[12] It is a powerful quantitative and qualitative analytical tool for elemental analysis of materials such as plaster and mortar. It is used to analyse major elements such as silicon dioxide (SiO_2), titanium dioxide (TiO_2), aluminium oxide (Al_2O_3), ferric oxide (Fe_2O_3), manganese(II) oxide (MnO), magnesium oxide (MgO), calcium oxide (CaO), sodium oxide (Na_2O), potassium oxide (K_2O) and phosphorus pentoxide (P_2O_5).[13] In masonry heritage building conservation in Malaysia, XRF test is conducted to identify the ratio of aggregate (silicon dioxide), calcium oxide and other elements.

3.4 Ion Chromatography Test

Ion chromatography (or ion-exchange chromatography) test is a chromatography process that separates ions and polar molecules based on their affinity to the ion ex-changer.[14] In building conservation, ion chromatography test is usually conducted to investigate the level of salt embedded in the wall. The three types of destructive soluble salts commonly found in the masonry wall of heritage buildings are chlorides (Cl), nitrates (NO_3) and sulphates (SO_4). These soluble salts can be detected by conducting ion chromatography test.[15] The test is conducted in three stages which are before salt desalination treatment (before cocoon method), during salt desalination treatment (after application of first layer of cocoon poultice) and after salt desalination treatment. The samples are collected from particular spots on masonry walls by drilling at different depth (same spot): 10mm, 20mm and 40mm. Ion chromatography test is important to determine the effectiveness of salt desalination treatment. The minimum salt safety level is 0.5%.

3.5 Moisture Test

Moisture is one of the key factors of building defect. There are various instruments or devices for measuring moisture level at heritage buildings. In general, there are three most common moisture detecting devices such as electrical conductance moisture meter, the carbide 'Speedy' meter and the hygrometer.[16] In Malaysia, moisture meters are used to detect the existence of dampness and moisture in masonry heritage building to ensure the effectiveness of the conservation works.

3.6 Schmidt Hammer Rebound Test

Schmidt hammer rebound test is conducted for non-destructive testing of the quality of concrete in the finished structure. This test provides an approximate indication of the compressive strength of hardened concrete.[17] It is also used to test and analyze the compressive strength and level of porosity of wall plasters and bricks. In building conservation in Malaysia, Schmidt hammer rebound test is also used to identify the compressive strength of the new lime plaster mock ups before re-plastering. The new lime plaster mock ups should be left on the building exterior wall that is exposed to sunlight and rain water for at least one month before carrying out the Schmidt hammer rebound test.

3.7 Function of the Scientific Tests

Table 1. The Function of the Scientific Tests

No.	Scientific Test	Purpose/ Function
1	X-ray Fluorescence (XRF) Test	-To identify the existence of the elements in a sample through elemental analysis. -To determine the contribution percentage of the elements in a sample. -To determine the percentage or ratio of aggregate, calcium oxide and other elements in the existing wall plaster and mortar.
2	Ion Chromatography Test	-To investigate the level of salt embedded in the wall, column and other building components and elements. -To determine the effectiveness of salt desalination treatment.
3	Moisture/ Dampness Test	-To investigate the moisture or dampness level in the wall, column and other building components and elements.
4	Schmidt Hammer Rebound Test	-To test and analyze the compressive strength and level of porosity of wall plasters and bricks. -To determine the compressive strength of the new lime plaster mock ups before re-plastering works carry out on site.

3.8 The Importance of the Scientific Tests

Scientific tests such as X-ray fluorescence (XRF) test, ion chromatography test, moisture/dampness test and Schmidt hammer rebound test are important to ensure the quality of masonry heritage building conservation from the aspects of building material composition and strength, accuracy of building diagnostics and effectiveness of conservation methods.

Table 2. The Importance of the Scientific Tests

No.	Scientific Test	Aspect	Importance
1	X-ray Fluorescence (XRF) Test	Building material composition	To minimise the risk building defects such as cracking or failure of wall lime plaster.
2	Ion Chromatography Test	Effectiveness of conservation methods	To determine the effectiveness of salt desalination treatment by comparing the salt level (before and after treatment).
3	Moisture/ Dampness Test	Accuracy of building diagnostics	To determine the areas or locations of walls, columns and other building components and elements with high moisture or dampness level. It is important to minimise the intervention in heritage building conservation.
4	Schmidt Hammer Rebound Test	Building material strength	To ensure the quality and compressive strength of the new wall plaster used for re-plastering works of heritage building.

3.9 Basic Guidelines to Conduct the Scientific Tests

Table 3. Basic Guidelines to Conduct the Scientific Tests

No.	Scientific Test	Basic Guidelines
1	X-ray Fluorescence (XRF) Test	-Collect samples of the existing lime plaster or lime mortar on site. At least three samples for comparison purpose. -Minimum 50 gram per sample. -Do not use metal or concrete made tools during the sample collection process as it may affect the result. -Record the location of the samples collected on the architectural drawing and report for documentation purpose. -Send the samples to the laboratory for testing. -Analyse the test result to identify the ratio of aggregate, calcium oxide and other elements for further action.
2	Ion Chromatography Test	-Collect samples from particular spots on masonry walls by drilling at different depth (same spot): 10mm, 20mm and 40mm. -The test shall be conducted for two stages which are before and after the salt desalination treatment process. -Do not collect the samples during the raining day as it may affect the result. -Record the location of the samples collected on the architectural drawing for documentation purpose. -Send the samples to the laboratory for testing. -Analyse the test result to identify the level of salt embedded in the walls for further action.
3	Moisture/ Dampness Test	-Collect data of moisture/ dampness level (internal and external walls and structures) with the aid of moisture meter. -Mark and label the level of moisture or height of rising damp on the walls and structures for reference and further action. -Record the moisture/ dampness level data on the architectural drawing and report for documentation purpose. -Analyse the test result to identify the level of moisture/ dampness for further action.

4	Schmidt Hammer Rebound Test	-The plunger of the Schmidt hammer is pressed against the spot of the selected surface to be tested. -Collect 16 readings in grid arrangement at a surface of 1 square foot. -Note: For new plaster mock ups, the mock ups shall be left at site and exposed to sunlight and rain water for at least one month before carrying out the Schmidt hammer test. -Calculate the mean- R value, standard deviation and estimated cube compressive strength (N/mm^2). -Record the compressive strength data in the report for documentation purpose. -Analyse the test result to identify the compressive strength of the wall plasters for further action.

4 Conclusion

As heritage buildings have significant cultural and historical values to the community and country, the application of appropriate conservation methods or techniques is the key to safeguarding the nation's heritage. The decision making tool in heritage building conservation projects should be evidence-based practice as risk always involves an element of uncertainty. Hence, scientific tests such as X-ray fluorescence test, ion chromatography test, moisture/ dampness test and Schmidt hammer rebound test should be conducted in all masonry heritage building conservation projects in Malaysia to ensure the quality of masonry heritage building conservation in the aspects of building material composition and strength, accuracy of building diagnostics and effectiveness of conservation methods. Without application of the scientific tests, the quality of the masonry heritage building conservation and the cultural significance of the masonry heritage building in Malaysia will be at risk.

References

1. A. A Ghafar, A.R. Haris Fadzilah. *Treatment of Salt Attack and Rising Damp on Heritage Buildings in Penang, Malaysia*. J.C.D.C., **15(1)**, pp.93-112. (2010)
2. B. Paulo Lourenco, et al. *Characterization and damage of brick masonry*, Springer, Historical and Masonry Structures Group, University of Minho, Portugal. (2014)
3. N. Bianco et al. *Hydraulic mortar and problems related to the suitability for restoration*, J. Periodico di Mineralogia (2013), **82**, 3, pp.529-542. (2013)
4. A. A Ghafar, A.R. Haris Fadzilah. *Restoration and Refurbishment of Old City Hall, George Town, Penang, Malaysia, Proceedings of 2005 International Conference on Construction and Real Estate Management, 2005, Wang Yaowu eds.*, China Architecture and Building Press, Vol. **2**, pp. 1356-1361.(2005)
5. D. Young. *Technical Guide Salt Attack and Rising Damp, a guide to salt damp in historic and older buildings*, Heritage Council of NSW; Heritage Victoria, Department of Planning and Community Development; South Australian Department for Environment and Heritage; Adelaide City Council. (2008)
6. R. Ana Pereira, James, D. *Interventions In Built Heritage: Managing Which Risks And For Whom, CIB W070 Conference in Facilities Management, Heriot Watt University, Edinburgh, 2008*, pp.25-32.(2008)
7. BSI. *Risk Management vocabulary – Guidelines for use in Standards*, London: British Standards Institution, Guide 73. (2002)
8. J. Douglas, Ransom, B. *Understanding Building Failures*. 4th ed. Oxon: Routledge. (2013)
9. G. Bowles, JR. Kelly. *Value and Risk Management*, D19CV9 Course Notes for Msc. Construction Management Programme, Heriot-Watt University, Edinburgh. (2005)
10. J. Douglas, B. Ransom. *Understanding Building Failures*. 4th ed. Oxon: Routledge. (2013)
11. British and World English. *In: Oxford Dictionaries* [online]. England: Oxford University Press. Available at https://en.oxforddictionaries.com/ definition/ scientific_method (2016) [Accessed 24 October 2016].
12. Amptek.com. *Overview of X-Ray Fluorescence Analysis*. [online] Available at: http://amptek.com/ xrf/(2017) [Accessed on 31 March 2017].
13. K. Wirth, A. Barth. *X-Ray Fluorescence (XRF)*, Sc. Edu. Resource Center, Carleton College, USA. (2016)
14. Wikipedia. *Ion Chromatography*. [online] Available at: https://en.wikipedia. org/wiki/Ion_chromato graphy (2016) [Accessed on 11 November 2016].

15. A. A Ghafar, A.R. Haris Fadzilah. Treatment of Salt Attack and Rising Damp on Heritage Buildings in Penang, Malaysia. *J.C.D.C.*, **15(1)**, pp.93-112. (2010)
16. J. Douglas, B. Ransom. *Understanding Building Failures.* 4th ed. Oxon: Routledge. (2013)
17. J. Douglas, B. Ransom. *Understanding Building Failures.* 4th ed. Oxon: Routledge. (2013)

Architecture typology of a British-era colonial schools at the World Heritage Sites of Georgetown and Melaka: Impact on heritage building conservation sector

Muhammad Adam Che Yusof [1 *], and *A Ghafar* Ahmad[2]
[1,2]School of Housing, Building and Planning, Universiti Sains Malaysia

Abstract. Colonial schools are among the most valuable and precious treasures from the British administration era. The architectural characteristics of the schools contain a wealth of meaning and philosophy in each of the building details that is not found in modern schools nowadays. However, in this era of rapid development, the status of colonial schools is unclear in terms of their heritage status and significance towards society or even the authorities. Moreover, some colonial schools that are now overshadowed by new buildings that were built inside of the school compound and its surrounding. Besides, the local authorities themselves are lacking on the documentation of colonial schools in terms of their value, and the principles behind the architectural style of the colonial schools that could prove beneficial to many parties later as a reference. This article will outline the method to handle this issue besides suggesting a relationship between the value of the colonial architecture and its history at both heritage sites of Georgetown and Melaka. Besides that, we will also classify different colonial schools according to their architectural style. To ensure the objectives are achieved, qualitative methods will be applied including several approaches such as descriptive method, historical method and content analysis method. Hence, this research can serve as a reference point and documentation, especially for conservation purposes of colonial schools. In addition, the local authorities can also improve their Conservation Management Plan (CMP) by adding a colonial schools sector for conservation work and later guidelines. This research will hopefully also encourage the younger generation on the importance of skills and knowledge in the heritage building conservation sector.

Keywords: Building Conservation, Typology, Colonial School, Architecture Style

1 Introduction

Malaysia had been colonized by several powers such as the Portuguese, Dutch and British in the past. These colonisers had their own cycle of development and downfall during their administration in Malaya [1]. In the process, several developments were brought to the country in terms of economic, education, trading and politics. As we look into the development and architectural part, it is clear that the colonisers had better expertise than us at the time. The colonial buildings that had been built are the most valuable properties to the country and were adapted from each colonisers' own architectural style albeit modified to suit the climate and context of this country, which is hot and humid. As such, the buildings demonstrate the combination of several of culture influences such as Chinese and Indian as they immigrated to Malaya [2]. This is clearly shown in the uniqueness of architectural style of colonial buildings and its motifs that contains heritage value that should be preserved until today. The term colonial buildings usually refers to shop houses, places of worship and government offices which are more popular among society. However the schools that had been left over from the colonial era should also be recognised as historic buildings that need to be preserved to maintain and protect the historical and aesthetic values of the buildings.

2 Background of the study

In the field of historical building conservation, there is a variety of approaches that can be appled. The understanding of each conservation work approach is very important to prevent any confusion during the conservation works being carried out [3]. Furthermore, it is important to ensure the originality, authenticity and art of the historical buildings is well preserved and not affected by any circumstances since the historical buildings differ from each other in terms of architectural style, structure, function and also types of defect so that the conservation works that need to be applied will also be different, based on the type of building [4]. This article will focus on the assessment of the different colonial school buildings based on their architectural style, building façade, building material, schools' site and also the arrangement and division of spaces in the buildings and their surroundings. In addition, this article will also attempt to link the history of the schools with the elements of architecture while highlighting methods to solve the issues and problem. The term "Typology" carries a broad meaning in categorizing certain object or figure physically to expose the values concealed behind the physical structure, not to mention its link with history. In Malaysia, research related to typology is rarely published especially if it is related to colonial schools in the World Heritage Sites. Most of the previous research on colonial buildings are more focused on shop houses in these historical cities. Hence, with the establishment of this kind of typology research it could indirectly contribute and impact on the building conservation sector in Malaysia ; at the same it will hopefully provide a valuable resource to many other parties such as academicians, historians, school management or even the local authorities for various other uses in the future.

Table 1. Types of typology for this research paper.

Typology	Explanation
Facade	To study the design that have been used to produce a façade that resembles its own architectural style during construction such as window design, door, wall ornament and column. [5,6]
Material	Identify the type of material used during the construction, such as the use of basic materials; brick, building roof, paint. [6]
Plan	Understanding the interior layout plan of the building and also the actual location of the colonial school; main road, back lane and the interior spaces. [5]

3 Problem Statement

To date, there are still many historical building in this country that are still facing defects whether to their façade, structure or else have been renovated without any proper conservation works. There are some buildings that have been abandoned by their owners and not well maintained due to lack of knowledge about conservation matters and also no initiative by the owners to preserve the buildings [7]. Based on observation, there are some old buildings that have lost their authenticity in their design, building material and architectural style due to the lack of proper maintenance and also the huge scale of renovation conducted on the building [8]. The weaknesses in maintaining such heritage treasure had caused serious defects to the structure and also the building fabric. Furthermore, based on a study by Raja Nafida Raja Shahminan (2008) there are some of the interior layout of shop houses in Melaka that had been changed by the building owner without any guidelines. In the case of this article, the researchers have selected the typology for the colonial school, and clearly there are many colonial schools nationwide, including those found in George Town and Melaka. These schools were used since inception until now without any change of use. Colonial schools have tourism potential and could attract academician to study more about the school's history, its architectural design or maybe as reference in developing new schools provided these colonial schools are well maintained and preserved with proper conservation methods.

3.1 Lack of documentation from local authorities.

In the Special Area Plan (SAP) of George Town and Melaka, there is a lack of specific and detailed information about colonial buildings or precisely the colonial schools in both heritage cities. Based on the researchers' observation, authorities such as Heritage Conservation Department, Pulau Pinang City Council (MBPP) and Building Control Department, Melaka Historical City Department (MBMB) agreed that for now they are lacking in terms of detailed documentation regarding the colonial school either from the aspect of pattern or architectural design, structure and the building materials or even the influencing of the interior layout pattern and the colonial schools' façade.

3.2 Neglected heritage status of colonial school.

According to the online news portal Mstar.com.my on October 2016, Penang Free School once planned to change the name of the school to Sekolah Menengah Kebangsaan Penang Free School. However, Raja of Perlis, Tuanku Syed Sirajuddin Syed Putra Jamalullail who is also one of its distinguished alumni insisted to maintain the original name of the school as the school is renowned as the oldest English Stream School in Malaysia and South East Asia. Besides this, there are also other cases involving schools in Seremban namely SMK Convent that was demolished for a new water reservoir. Similarly, SJK(T) Convent Seremban was also demolished and new school buildings were rebuilt. This shows that the sentimental values of the old school are ignored in the eyes of some of our community who fail to make the effort to preserve and take care of these colonial schools.

3.3 Low quality of the school building.

The façade quality of the colonial schools nowadays are declining as the buildings are aging year by year. In this case, the schools management and local authorities have to ensure that any related issues connected with the buildings should not be overlooked so that in the future it will not harm the building further. Moreover, the schools management should not change the façade or any part of the building by themselves without adhering to appropriate guidelines and permission from the local authorities. This is to ensure that all renovation works that are planned are based on the prepared guidelines so that the conservation work will be in line with the conservation act.

4 Research Area

In Malaysia, there are several places and cities recognized by UNESCO as World Heritage Sites such a George Town and Melaka. The Asian and European influences in these heritage cities are clearly shown by their historical buildings such as the places of worship, government buildings, shop houses and fortresses. Both cities are recognized by UNESCO on 7th July 2008 and had been chosen as researcher's scope of study area [9].

4.1 George Town, Pulau Pinang.

George Town is one of the fastest and rapidly developing cities in Malaysia located 325 kilometres north from Kuala Lumpur. The city that was built at foothills was discovered by the English trader, Sir Francis Light in 1786 and at that time there was no development or artistic buildings because at first it was not intended to become a settlement [10]. Since the 1900s, many buildings were built such as the administration offices, housing, places of worship – mosques, churches, temples – and schools. Those buildings adopted the combination of local architecture with several of foreign influences such as English, Arab, China and India making them until today one of the examples of precious heritage architecture that needs to be protected. Furthermore, as George Town is one of the historical city in Malaysia, it is divided into two zones whereby the core zone consists of 109.38 hectares while the buffer zone encompasses 150.04 hectares [11].

4.2 Melaka Historical City, Melaka.

Melaka Historical City is the capital city for Melaka State located 125 kilometre south of Kuala Lumpur and was the 8th city that existed in South Peninsular Malaysia. Located near the estuary of Melaka River that flows towards the Straits of Melaka, this city contains thousands of history making it well developed today beginning from the Melaka Sultanate Era until the colonization of Portuguese, Dutch and British [reference 9]. Consisting of two zones, Melaka covers 242.8 hectares of core zone which is 83.4% of the city and 45.3 hectares of buffer zone that consists of 15.7% of the city area.

Table 2. Colonial schools at both heritage cities.

Melaka	Age	George Town	Age
SMK & SK St Francis	1880 (137 years)	St. Xavier Institution	1852 (165 years)
SK Banda Hilir	1908 (109 years)	SMK & SK Convent Light Street	1852 (165 years)
SK Sacred Heart Convent	1905 (106 years)		
SMK Tinggi Melaka	1826 (191 years)		

Based on the table above, the schools are selected based on the criteria as follows:

i. Located in the area of UNESCO World Heritage Sites: core and buffer zone.
ii. Aged 100 years old and above.
iii. Implemented the colonial architecture style.
iv. Only heritage buildings in category I and II.

5 Objectives

The overall objective of this research is to know and identify the typology of the colonial schools' architecture style while drafting the method and approaches to address the related issues.

5.1 Drafting methods to address the issues and problems related to the maintenance of colonial school buildings at the George Town and Melaka World Heritage Sites.
This objective is to research and develop the method from several parties to overcome the related issues regarding the colonial schools buildings.

5.2 Classify each different colonial school according to the school's architectural style.
Classifying each different school according to their architectural design and principle that had been applied based on the given typology in this article.

5.3 Establishing the relationships between colonial architecture and school history at George Town and Melaka World Heritage Sites.
To identify the relationship between architectural design of the buildings and the history of the school.

6 Methodology

The research applies Qualitative method in this study with several approaches. The type of approaches in the data collection of this study are as follows:

6.1 Observation : Applying the structured observation means that researcher needs to identify the aspects required in the method such as the types, patterns or the style of the buildings. The observation method will be recorded through the prepared inventory form and also through pictures captured on the site.

6.2 Document Analysis: This approach is required to obtain additional data and resources other than interview and observation since it is the fastest way to collect more references. It is use to obtain the validity in the research that also can shaped the guidelines of the research. For example, historical approach entails document analysis method that can be used to seek the connection between the history and the condition of the building. Historical approaches will also lead to the related issues from the historical observation and determines from which angle the research could be done [5].

6.3 Interview : A technique of collecting data by interviewing the research subjects. In this research, semi-structured interviews are used which means the questions are already prepared by researcher but the arrangement and the sequence can be modified [12]. There may also be additional questions, or some of the prepared questions may even be skipped during the interview session.

6.4 Case Study: This approach helps researchers to select the suitable case study for the research based on the specific criteria that have been chosen to achieve the goals of the study. For instance, this study identified several criteria to select the best schools to be included, such as the schools must be within World Heritage Site boundary, built in colonial era and categorized as Category I and II Heritage Buildings.

7 Discussion

For this study, the findings might be useful for many parties as a reference or record regarding the particular schools. Moreover, it also will benefit the construction sector such as the conservator, contractor and local authorities since there are prepared documentations containing the architectural elements of the school if there is any conservation work that needs to be worked out. The findings of this study will explain the typology of the selected schools based on their types of typology which can be seen from their façade, material and plan. It will also describe the architectural style and patterns of the schools which make it easier to categorize and differentiate between school's buildings. On the other hand, some limitations include failure to study some of the schools because the land owner refused to grant permission, and also the lack of the cooperation from several parties; local authorities, school's management and consultants, and the schools are restricted areas rendering them out of bounds without proper permission.

8 Impact on Conservation Sector

8.1 Improvement on Conservation Management Plan (CMP)

It will help the planning from local authorities in managing the heritage city from the aspect of conservation buildings especially in the context of colonial schools. In this case, this article could help to categorize the colonial school in their own specific category without needing to be categorized under one big group which is in category I and II buildings.

8.2 References for several parties

The typology of the schools are very important as they show the detailed elements that were hidden under the schools buildings. Academicians, historians and conservator can use this typology documentation as their references. On the other hand, schools management and the ministry for example could make this typology paperwork as a reference and archive document for future use.

8.3 Enhancing the quality of conservation works.

This is shown when the conservation works for those colonial school are running, so that the detail documentation about the schools are already prepared whether from the history aspect until its architectural concept. For instance, conservator and contractor that are planning to run the conservation works could refer the building material used by the schools for them to get the actual material from any related supplier based on the selected criteria without breaking the guidelines. It is also make the conservation works easier when the detailed documentation are already prepared especially when it comes to the building typology as it act as a measurement to the conservation works procedure.

9 Conclusion

As a conclusion, those colonial school building contains the different layout of space, design and also the material use from one another. Therefore, to run the conservation works on those buildings, the typology aspect need to be carried out and not only focusing on the preservation on the building facade. The typology aspect that need to be carry out are the facade, plan and material typology. Typology study not only can be done for colonial school building but it is also can be done to any historical building such as the Malay houses, museums, clock tower or even fotresses. In addition, Malaysia and local authorities already have their own guidelines and Akta Warisan Kebangsaan 2005 that are still in use, there are still some certain parts and section that need to be impove for the benefit of conservation works management. The earlier planning also important to determine the appropriate method to solve some issues so that the correct technique can be apply to the particular buildings and ensuring the authenticity is well preserved. In fact, with the existence of typology for those colonial schools may also reminds the early moments of the schools were built hence it will reinforces the historical elements in the school building.

References

1. A. S. Hassan, S.R.C Yahaya, *Architecture and Heritage Buildings in George Town Penang* (2012)
2. http://zahirzainudin.blogspot.my/2014/06/kepelbagaian-seni-bina-dan-struktur.html
3. A. A. Hussin, *Pemuliharaan dan pemeliharaan warisan di Malaysia*.(2011)
4. A. G. Ahmad Rangka kerja pemuliharaan bangunan warisan di Malaysia. Bengkel Konservasi Monumen dan Tapak Tanah Bersejarah, 27-29hb. November 2006, Melaka. Kerajaan Negeri Melaka dengan kerjasama Perbadanan Muzium Melaka (PERZIM), Melaka.
5. R. N. R. Shaminan, *Kajian Tipologi Rumah Kedai Awal Era Belanda di Bandar Melaka : Sumbangan Kepada Bidang Pemuliharaan di Malaysia* (2008)
6. http://tutirahayu-knowledge.blogspot.my/2012/05/tipologi-arsitektur.html
7. P. Keromo, Faktor Kerosakan ke atas Monumen dan Tapak Tanah Bersejarah, Bengkel Konservasi Monumen dan Tapak Tanah Bersejarah, Melaka 2000, (2002).
8. A. G. Ahmad, *Pemuliharaan Bangunan Warisan di Malaysia: Pengalaman dan Cabaran Masa Hadapan* (2010)
9. A. G. Ahmad, *Pemuliharaan Bangunan Warisan di Malaysia: Pengalaman dan Cabaran Masa Hadapan* (2010)
10. N.W.Queen, L.Ayob, *Early History of Penang* (2012)
11. Jabatan Perancangan Bandar dan Desa, Rancangan Kawasan Khas George Town (2013)
12. S. N. Harun, *Amalan Kerja Pemuliharaan Bangunan Bersejarah di Malaysia* (2005)

Benefits and Aura of Rubies and its Local Intelligence among the Malays Community

Noor Azam Azmi[1] , *Kai Xin* Tay [2], *Badaruddin* Mohamed[3] & *Salasiah* Che Lah [4]

[123] School of Housing, Building & Planning, Universiti Sains Malaysia
[4] School of Humanities,Universiti Sains Malaysia

Abstract. Rubies are associated with aura and nutrients. Many scientifically researches have been carried out for examining the rubies as materials, however, the little-existed literature on beliefs and usefulness of rubies, especially Malay community who have used Rubies for many decades. Therefore, this paper is to critically review the benefits and aura of rubies from Malay local intelligence. This paper found that the Malay community is knowledgeable on the benefits of rubies. Characteristics that exist in the stone itself have been identified that this gem is a class of its own. The beliefs that Malay community has on rubies can be associated with local knowledge that exists about the gem are also discussed

1 Introduction

The Malay community is synonymous with the use of rubies as accessories, for example, rings. There are many types of precious stones such as Akik, Opal, Turquoise Stones, Jade, Amethysts, Pearls, Sapphire, Topaz, Emeralds and much more. Each and every stone mentioned possessing its own benefits and aura (Karyasa et al., 2015). Besides being exceptionally lovely, these precious stones are also being kept as special collections among the Malays (Gimlette, 1920). The influence that these stones portray has existed past decades (Ball, 1935) and the Malay community uses them in many different ways, such as medical. Apart from being used as accessories, these stones are also associated with the symbol of power, wealth, love and also zodiac (Ball, 1935). Generally, the Malay community uses precious stones as amulets to protect themselves from negative elements, such as devils and demons, to keep diseases at bay and also to maintain health (Gimlette, 1920). Merchants and businessmen are frequently seen to be selling the precious stones in places where the Malays dominantly visit such as night markets, expo, shopping complexes and so on.

Rubies are the focus of this study. Generally, rubies tend to be red in colour (Park et al., 2007; Ronald W. Arends, 2000). Many facts and beliefs among the community toward rubies. Rubies are widely known as a red colour, which is the symbol of love, power and pride. The title 'king of gemstone' was given to justify that this particular stone is famous not only among the Malays, but also throughout the world (Bank et al., 1988; Joseph, Lal, Shinde, & Padalia, 2000; Park et al., 2007; Spiegehan, 2000). Rubies become synonymous with the Malays due to its physical beauty, aura and the benefits possessed. According to Flöter, Lujala, and Rød (2005), rubies are also in pink and dark red. The Mohs scaling[1] of rubies is 9 and this proves that rubies are hard and strong (Joseph et al., 2000; Sioshansi, 1992). The best rubies are found in Myanmar (Bank et al., 1988) and good quality rubies are found in Tanzania, India, South Africa, Pakistan and Afghanistan. Rubies are assumed to be the birthstone for those born in the month of January and those who fall under the zodiac sign of Capricorn (Yeni Fitrianingsih, 2014).

[1] A scale of hardness used in classifying minerals. It runs from 1 to 10 using a series of reference minerals, and a position on the scale depends on the ability to scratch minerals rated lower.

2 Characteristics of Rubies

Ruby, also known as red rubies, is a type of red stone and its colour ranges from light red to dark red (Kane, 1984). Rubies are a type of corundum minerals (A1203) (Calligaro, Poirot, & Querré, 1999; Herman, 2008; Kane, 1984). Besides being physically attractive, rubies are also durable due to its high Mohs scale[1] (9 to 10). The red on rubies is the effect of chromium element, which possesses a high content of aluminium oxide and crystal hexagonal material (Joseph et al., 2000; Stone & Vickerman, 1971; Vickerman, 1971). The Malays named rubies as Delima due to the colour and shape which resembles pomegranates. The Malays regard rubies as one of the most precious stones beside sapphire, emerald and diamonds (Gimlette, 1920). The characteristics of popular rubies among Malays are presented in Table 1.

Table 1: The characteristics of popular rubies

Delima Darah Merpati (Pigeon's Blood Rubies) It is known for dark red in colour with some infusion of blue. This precious stone has the highest value in the market with high demand.	*Delima Bintang* (Star Rubies): Possess high quality and is able to reflect light. If the stone is projected with a light beam, this ruby is able to emit red light.

The price of the rubies is determined according to its colour. The most expensive and valuable stone should be sparkling in red (Ball, 1935). The clarity of rubies is also another characteristic, which determines its price and value among the Malays. Due to the close bond rubies share with the Malay community, certain beliefs and myths about this stone exist.

3 Beliefs on the Benefits and Aura of Rubies in the Malay Community

The Malay community believes that rubies have benefits and safeguards a person from threats such as black magic and from dignitaries and rulers. Individuals wearing rubies are said to be guarded against mystical disturbances. Rubies were worn by ancient Malay dignitaries due to Malay believes it could trigger majestic aura, whereby peasants and servants would dutifully obey, respect and bow to their commands (Gimlette, 1920).

The effects towards the servants would definitely be the ideal dream to the rulers in the sense that it will be easier for them to control and lead a group of people without any rebellion and dissatisfaction towards the leaders. Dignitaries in the old era not only used rubies in their rings but also pinned the stone into their kris. As for the merchants, they believe that rubies are able to bring benefits to their businesses. Owing to this fact, it is not surprising that most Malay merchants and businessmen possess much interest in owning ruby-stoned rings.

The Malay community believes that original and pure rubies are sanctified as this type of stone is associated with the mystical world. It is believed that a person who owns rubies are no ordinary individuals. It is supposed that for one to own rubies in the olden days will need to go through certain rituals and special ceremonies. Ancient Malay community, especially the Javanese Malay believed that rubies were originally genies which turned into a stone (Triyogo, 2005) and for one to possess a pure stone, the person needs to have a clean soul. If an owner would like to switch the possession to another person, it has to be voluntarily done without the involvement of money or other exchanges. Once sold, the rubies lose their benefits. The Javanese used to mention that owning rubies are usually fated and destined (Triyogo, 2005). The Malay community, especially in Indonesia, highly believes that anyone who owns authentic rubies will be invulnerable and impenetrable to sharp weapons (Triyogo, 2005). It is believed that no one can cut the hair or stab the skin of a person who wears a ring made from rubies.

4 Local Malay Intelligence on Rubies

There is various local wisdom related to the rubies among the Malay community especially knowledge inherited from their ancestors (Gimlette, 1920). Firstly, rubies can be used in healing processes and medical purposes. There are a small group of Malays who still use rubies as a tool to heal diseases. For instance, sickness includes joint pain, cardiovascular diseases, nervousness, insomnia, sinusitis, high blood pressure, diabetes and many others. One way healing those diseases mentioned is through the use of rubies. The first method used in healing is by placing rubies in or around pain areas. In the second method, the rubies would be smashed into powder form to be consumed by patients. The former method treats external pain while the latter focuses on healing internal pain. It is also said that the presence of rubies alone may be able to protect an individual from getting ill. To maintain youngness and health, it is advised to soak rubies in a glass of plain water for about an hour and to drink the water thereafter. Authentic rubies will turn the plain water into the red.

Second wisdom is the methods to determine the real and high-quality rubies. A way to determine if the rubies are of high quality is by looking at the visual form of the stone. Stones of good quality can be seen to be having fibres or thin thread-like strands (Herman, 2008). There are some apparent differences between original rubies and synthetic ones. Fake rubies are appeared to be lighter than the original ones. In addition, real rubies are cooler when touched compared to the fake rubies, due to fake rubies are usually made from glass and other synthetic materials. Natural rubies are similar in size to pomegranate seeds, which is moderate in size. Another method to test the authenticity of rubies is to soak them in a glass of water. The water is then continued soak into another seven glasses of water and if the water still turns red colour, the rubies are considered pure and of high quality. The red colour of water would be visible for a one-month period. On the contrary, the red on synthetic rubies can easily fade as they contain phosphor. It is also said that a glass containing authentic rubies will be impenetrable even by bullets. The quality of rubies is also tested through their strength whereby an original piece will stay intact even when hammered. Besides, logics from Malay local intelligence suggest that natural and high-quality rubies may reach hundreds and thousands of ringgit in value. Any stone sold at a lower price may suggest that the rubies may not be original and are of low quality. However, this does not count in rubies which supposedly originate from the supernatural world that is prohibited from being sold.

Third, local intelligence on testing rubies that possess genies *(Khadam)*. The term '*khadam*' originates from the Arabic language which literally translates into a servant. In the context of local intelligence, these servants are known as a paranormal helper who lives in the supernatural world. The servants are frequently assumed to be genies. The existence of genies has been mentioned in the holy Quran. The servants are said to be creatures originating from the genies and angels (Kane, 1984). Theologians believe that the servants as mentioned in the holy Quran does not possess magical or supernatural powers as believed by certain Malay community and the idea that the servants residing in the rubies will make the stone magical has been brushed aside (Akhmad, 2013). The '*khadam*' mentioned in the Quran relates more to the protective system allocated to those who are duly followers and with high faith in Islamic teaching.

"For each one is successive [angels] before and behind him who protect him by the decree of Allah. Indeed, Allah will not change the condition of a people until they change what is in themselves. When Allah intends for a people ill, there is no repelling it. And there is not for them besides Him any patron". (QS. ar-Ra'd; 13/11)

The methods practiced by the Malay community in determining if a ruby possesses *khadam* as follow:

i. Free from impurities.
ii. Touch the ring using a right middle finger.
iii. The finger is then placed in the upper palate of the mouth.
iv. To recite the verse of An-Nas thrice with the eyes closed and pray with the intention to see the 'khadam'.
v. With the eyes still closed, place the thumb on the middle finger.

If the ruby possesses *khadam*, the following will happen:

i. Animal forms or colours will be visible
ii. Vibration and biting sensation will be felt on the middle finger

5 Discussion and Conclusion

As a Muslim who believes in the teachings of Allah, an individual should surrender fully to the Almighty God. No living or non-living creatures in the world is able to provide benefits unless with His permission. For example, the use of precious stones as rings is not prohibited but the benefits and aura projected from the stones is to be believed to have come from the permission of Allah S.W.T.

There is definitely a meaning behind every creation of Allah S.W.T. Thinkers and the intelligent would definitely look beyond every creation and investigate the unique creations. Rubies or any other materials are gifts and reward from Allah S.W.T with their own characteristics and nature such as the sweetness of sugar and saltiness of salt. Precious stones such as rubies possess their own specific benefits and aura and only those knowledgeable and following the right path will be able to reap the benefits over the matters mentioned. The higher content of metal minerals in a stone, the higher would be the electromagnetic waves formed. The effect of these waves on the human body is known as an aura (Fitrianingsih, 2014).

One idea to understand is the awareness that no living or non-living entities possess supernatural powers. In this context, the benefits and advantages associated with rubies are given and permitted by Allah S.W.T. Overall, dependence towards Allah has to be practised for one not to fall under the category of 'syirik' (polytheistic or believing that there is more than one God).

Acknowledgment

Much appreciation is extended to the Ministry of Higher Education and Long-Term Research Grant Scheme (LRGS), (Local Knowledge) with the account number 203/PTS/6727002, USM members and those involved in this study.

References

1. Akhmad, P. (2013). *Ilmu Hikmah Antara Hikmah Dan Kedok Perdukunan.* Adamssein Media.
2. Ball, S. H. (1935). A historical study of precious stone valuation and prices. *Economic Geology*, *30*(6), 630–642. doi:10.2113/gsecongeo.30.6.630
3. Calligaro, T., Poirot, J. P., & Querré, G. (1999). Trace element fingerprinting of jewellery rubies by external beam PIXE. *Nuclear Instruments and Methods in Physics Research, Section B: Beam Interactions with Materials and Atoms*, *150*(1-4), 628–634. doi:10.1016/S0168-583X(98)01004-0
4. Flöter, A., Lujala, P., & Rød, J. K. (2005). *the Gemstone Dataset Codebook.*
5. Gimlette, J. D. (1920). A Curious Kelantan Charm. *Journal of the Straits Branch of the Royal Asiatic Society*, *82*, 116–118.
6. H. Bank, Gübelin, E., Harding, R. ., Henn, U., Scarratt, K., & Schmetzer, K. (1988). An Unusual Ruby From Nepal. *The Journal of Gemmology*, *21*(4), 222–226.
7. Herman, D. Z. (2008). Pendayagunaan mineral untuk menjadi permata. *Museum Geologi*, *5*(7), 4–5.
8. Joseph, D., Lal, M., Shinde, P. S., & Padalia, B. D. (2000). Characterization of gem stones (rubies and sapphires) by energy-dispersive x-ray fluorescence spectrometry. *X-Ray Spectrometry*, (May 1999), 147–150. doi:10.1002/(SICI)1097-4539(200003/04)29:2<147::AID-XRS370>3.0.CO;2-K
9. Kane, R. E. (1984). Natural rubies with glass-filled cavities. *Gems & Gemology*, 187–189.
10. Karyasa, I. W., Dantes, I. G. R., Muderawan, I. W., Rai, I. W., Jasmani, J. P., & Rekreasi, K. (2015). Geokimia Batu Permata Kresnadana Pulaki, *4*(2), 604–610.
11. Park, G. S., Bae, S. C., Granick, S., Lee, J. H., Bae, S. D., Kim, T., & Zuo, J. M. (2007). Naturally formed epitaxial diamond crystals in rubies. *Diamond and Related Materials*, *16*(2), 397–400. doi:10.1016/j.diamond.2006.08.031
12. Ronald W. Arends. (2000). Enhanced Gem Stone And A Method Of Simulating The Appearance Of An Expensive Gem Stone. United State.
13. Sioshansi, P. (1992). Polishing process for refractory materials.
14. Spiegehan, J. M. (2000). A price above rubies: Patriarchy and the feminine. *Psychological Perspectives*, *40*(1), 70–78. doi:10.1080/00332920008403369

15. Stone, F., & Vickerman, J. (1971). Magnetic properties of chromium ions in oxide matrices. Part 1.—Cr 2 O 3-Al 2 O 3 solid solutions. *Transactions of the Faraday Society*, *67*, 316–328.
16. Triyogo, A. H. (2005). *Benda-benda bertuah masyarakat Jawa*. the University of Michigan: Narasi.
17. Vickerman, J. C. (1971). Magnetic properties of chromium ions in oxide matrices. Part 2.—MgCr2O4—MgAl2O4 solid solutions. *Transactions of the Faraday Society*, *67*, 665–673.
18. Yeni Fitrianingsih. (2014). *Pengaruh Hypnobirthing Terhadap Nilai Apgar Bayi Baru Lahir Pada Persalinan Normal Primipara*. Universitas Diponegoro.

Traditional *Kinulung* Craft Weaving of the Dusun Tindal of Kota Belud, Sabah

Salbiah binti Kindoyop[1], *Noria* ak Tugang[2], *Humin* Jusilin[3] and *Junior* Kimwah[4]

[1]Faculty of Applied and Creative Arts, Universiti Malaysia Sarawak
[21]Faculty of Applied and Creative Arts, Universiti Malaysia Sarawak
[3] Borneo Heritage Research Unit, Faculty of Humanities, Arts and Heritage, Universiti Malaysia Sabah
[4] Faculty of Humanities, Arts and Heritage, Universiti Malaysia Sabah

Abstract. Arts and cultural heritage is a potent reflection of the identity of a community. The art of traditional handicraft weaving in any society is dependent on the resources available in his environment and the needs of the particular community. The Dusun Tindal of Kota Belud is rich in weaving craft traditions with *kinulung* as one of its distinctive products. *Kinulung* is unique with its own specific purpose and simple design, and has been able to withstand modernisation and change. However the factors of adaptation, environmental change, and modern technological transformation could pose as threats to the traditional and heritage elements in the production of the *kinulung*. This study is important as a platform for the promotion of continuity and resilience of the Dusun Tindal heritage. Kinulung documentaton will clarify various aspects of the design and purpose elements of this traditional craft. The study focuses on the design and meaning presented in the *kinulung* weaving tradition as an object of culture in the community. Material for this study was derived from field work through observations, interviews and notations. The findings of the study indicate that the design and motif of the *kinulung* reflect the cultural values and local knowledge of the Dusun Tindal.

1 Introduction

This study on the *kinulung* craft aims to provide insights into an aspect of the cultural heritage of the Dusun Tindal, a majority ethnic group located predominantly in the district of Kota Belud, Sabah. Among the objectives of this study are: i) to present the local terms related to *kinulung*, and ii) to document the process and technique in the making of the *kinulung*. Basic materials used in this craft are *poring* (big bamboo, *gigantochloa levis*) and *tulu* (small bamboo, *schizostachyum brachyladum*).

In the past, *kinulung* was actively produced in villages around the Kota Belud district. However, although such craft is still produced nowadays, these no longer feature predominantly in the Dusun Tindal kitchen; the modern day *kinulung* serves a more decorative purpose rather than its original function as a food cover or food receptacle. There remains three ways to ensure the sustainability of this Dusun Tindal craft tradition: (1) by introducing the local terms and concepts related to the weaving of *kinulung*, (2) by the continuous production of the *kinulung* based on market demand, and (3) by documenting and publishing the aspects of design, technique and materials used for *kinulung* weaving. To ensure its continuity, the *kinulung* craft needs to be revitalised and made popular to the public, and the weaving techniques taught to the younger generation.

2 Methodology

Photography is used in practice–based research or other research situations for the acquisition of visual data, storage or management of visual information, analysis of visual data, and the presentation of research findings. Visualisation requires drawing, diagrams, concept maps or mind maps, flow charts, storyboards, matrices, and network displays which employ colour, tone line, plane, shape, scale, and symbol. Metaphors and analogies often serve to initiate or stimulate this process of visualisation. The interview is a key method in seeking the particular opinion of others about identified research topics or areas with a purposeful conversation initiated by the interviewer and focused on content specified by the research objectives.

Data for this study was derived from (1) notes, (2) photographs, (3) sketches, and (4) voice recorder. These techniques allowed the researcher to experience and acquire a critical visual understanding of the cultural practices relayed in the interviews and during observations.

3 *Kinulung* handicraft

The *kinulung* name originates from the word '*okulung'* which means the act of curving in Dusun Tindal. The curved shape of the *kinulung* is due to the material make- up which is appropriate for its function as a food cover or receptacle. The Dusun Tindal in the olden days would hang the *kinulung* within easy reach in the kitchen to keep or protect their food from insects, cats or dogs. During gatherings or celebrations, the *kinulung* becomes a tray for food presentation or preparation, its curved shape well suited as a receptacle for various food objects. *Kinulung* sizes vary according to the usage. For example, for food storage the size would usually be 40cm x 40cm. For food dishes, a larger size *kinulung*, around 60cm x 60cm, would be the preferred variant.

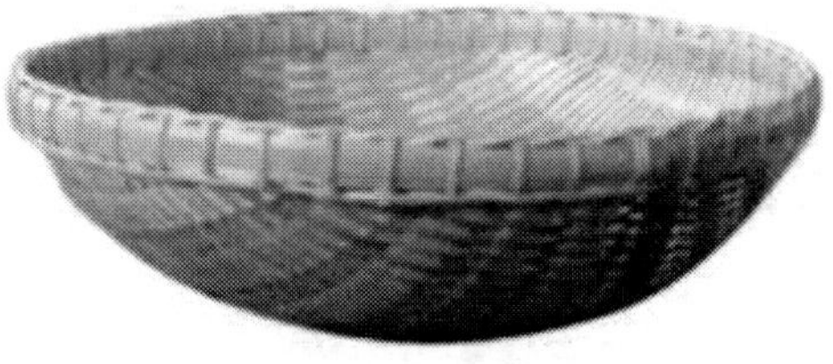

Figure 1: *Kinulung* design

Traditional kinulung is made from natural materials, usually *tulu*[1] *(schizostachyum brachyladum)* and *poring*[2] *(gigantochloa levis)* from the bamboo family, and *calamus javensis* rattan. The *tulu* is fashioned into small strips (*pinuhusan*) and the *poring* into wider versions (*pongorigi*) for the weaving process. *Kinulung* weaving adopts an overlapping technique with strips arranged alternately in vertical and horizontal positions. The *pinuhusan* forms the horizontal structure of the weaving while the vertical *pongorigi* serves as the mainframe structure (see Figure 2).

[1] *Tulu* - a type of small bamboo with a long trunk, thin outer layer and could easily be cut open.
[2] *Poring* - a large trunk bamboo suitable for use as mainframe in weaving as it holds its structure well.

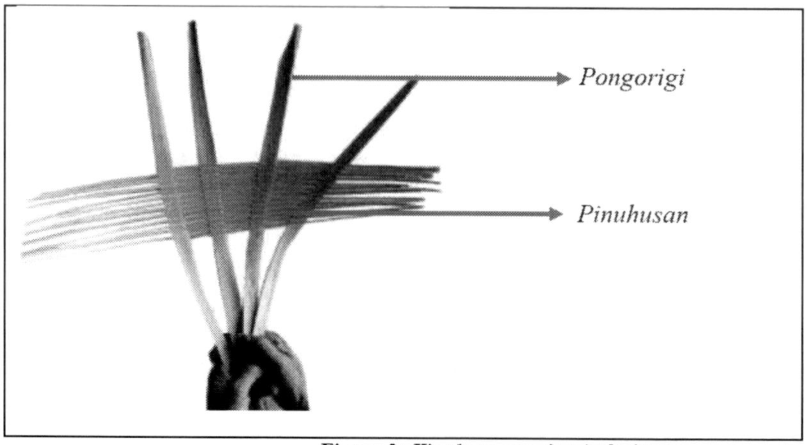

Figure 2: *Kinulung* weaving technique

1.1 Process and technique of *kinulung* weaving

Although *kinulung* weaving in itself is a fairly simple and easy technique, it is a laborious process. Weaving with small and thin bamboo strips requires skill and wisdom acquired over time or inherited from elders. The *mangatu* (weaving) process could be done any time of the day. Due to its small size, the *pinuhusan* is resistant to breakage or splitting. However the *pongorigi*, taken from the inner parts of the bamboo, must be moistened prior to the weaving process since it is more susceptible to damage or breakage. The supple *pongorigi* ensures the woven end-product is both smooth and neat.

Among the Kota Belud Dusun Tindal, the stages of *kinulung* production is defined according to the following sequence:

a) Mokiburuon - finding bamboo or suitable weaving material
b) Momutul - cutting the *tulu* and *poring*
c) Mamalapak - splitting and removing bamboo skin and inner part
d) Mongila - smoothening bamboo skin and inner part
e) Momuhus - trimming bamboo strips and *dulung* (frame)
f) Mangatu - weaving process
g) Monginsok - 'compacting' to ensure weaving is 'tight' (without gaps or space)
h) Momodulung - forming mainframe
i) Momolongkop - ensuring mainframe is neat and tied with rattan
j) Momolilit - weaving rattan on the edges of *dulung* or frame

 https://doi.org/10.1051/shsconf/20184502002

Table 1: Material and equipment used in making Dusun Tindal *kinulung*

Material	Method / Technique
(i) *Poring* (bamboo) 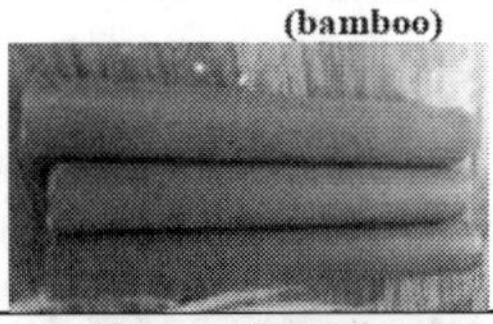	Bamboo is the main material used for weaving. The bamboo trunk is cut open and the inside and outside parts stripped off. The inner portion is trimmed into small strips.
(ii) *Sogoh* (rattan) 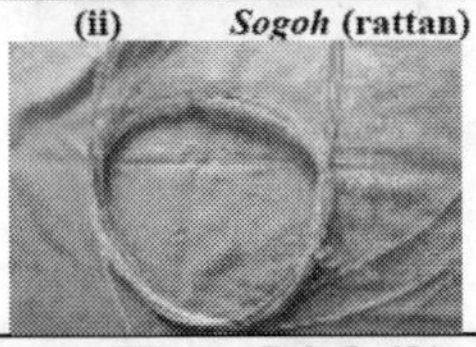	Used as anchor material or binder. Rattan is usually used for frame or frame trimming.
(iii) *Pais* (knife) 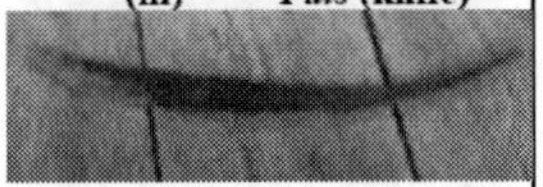	Used for cutting and smoothening bamboo strips and rattan.
(iv) *Dangol* (parang) 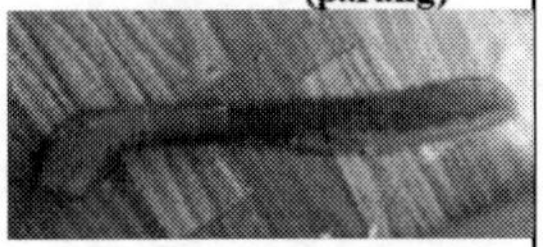	Used for cutting and splitting bamboo trunks.
(v) *Takang* (spanar) 	Used as the frame holder/ anchor during the *momodulung* process.

Table 2: Material and weaving techniqueof the Dusun Tindal *kinulung*

(i) Pongorigi 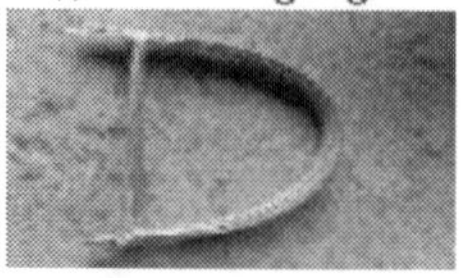	Large strips cut from the inner part of the bamboo trunk. Strips serve as pillars or boundary planes in the weaving.
(ii) Pinuhusan 	Smaller strips that have been trimmed and dried; these are woven or stacked on the *pongorigi* in alternate fashion from the outside to the inside or vice versa.
(iii) Tusak do sogoh 	Inner part of rattan that has been trimmed and made smooth. Used as a *pomolilit dih dulung* or as 'twine' to tie the frame neatly.
(iv) *Dulung* 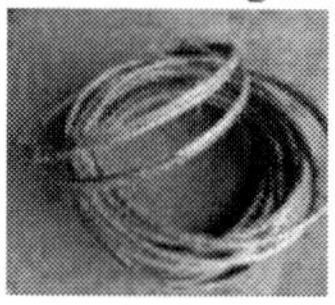	A frame made from the bamboo trunk which has been cut, made smooth, and trimmed into strips of the required thickness and width.
(v) Mongorigi 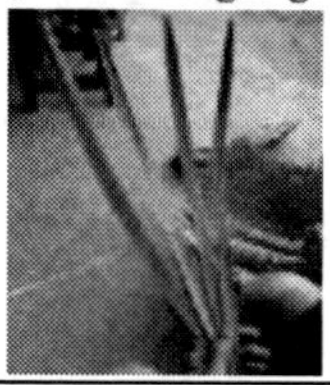	Start of the weaving process whereby three to four pieces of *pongoringi* are arranged in fan-like configuration; these act as the mainframe structure for the weaving.
(vi) Mamatu 	Weaving process which involves arranging *pinuhusan* and *pongorigi* in alternate sequence or layers.
(vii) Momodulung	*Momodulung* is the process of installing the outer frame to the completed woven piece.

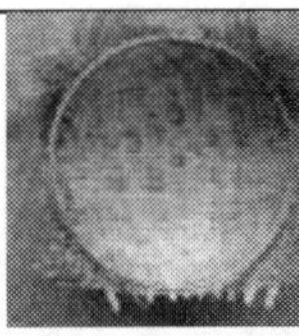	
(viii) Momolilit 	*Momolilit* is the process of tying the outer frame with rattan and ensuring the attachment to woven piece is neat and secure.

Table 2 outlines the local expertise in *kinulung* weaving material and technique. Four materials are essential in *kinulung* craft weaving: (1) *pongorigi*,(2) *pinuhusan*, (3) *tusak do sogoh*, and (4) *dulung*. These are utilised during the various stages of *mongorigi*, *mamatu*, *momodulung* and *momolilit*. According to the informants[3] of this study, *kinulung* weaving is comparatively easier for a skilled weaver as opposed to other crafts such as basket making since the materials and weaving techniques of the former are simpler and easily learnt/ acquired.

Dusun Tindal *kinulung* production involves three stages: (1) the preparation of weaving materials, (2) the weaving process, and (3) the installation of frame on the completed woven craft. The process of weaving itself has two methods: the first is by positioning a large strip (pongorigi) as the backbone and interspersing this with smaller strips (*pinuhusan*) with the sequencing arranged in such a way so as to produce the desired size and design. According to one of the informants,[4] this method could produce between 10 to 20 *kinulung* weavings in a day. However, the frame installation process takes a longer time since it requires more care and precision. A *kinulung* therefore might take about a week from commencement to completion.

Traditional *kinulung* weaving produces a horizontal and vertical line motif. The arrangement of bamboo strips in alternate order or sequence is presented in Figure 3(i). To obtain the *kinulung* curved form, the weaver will push and manipulate the materials by hand. The circumference of the *kinulung* frame has to be formed neatly to obtain the desired shape (see Figure 3(ii)).

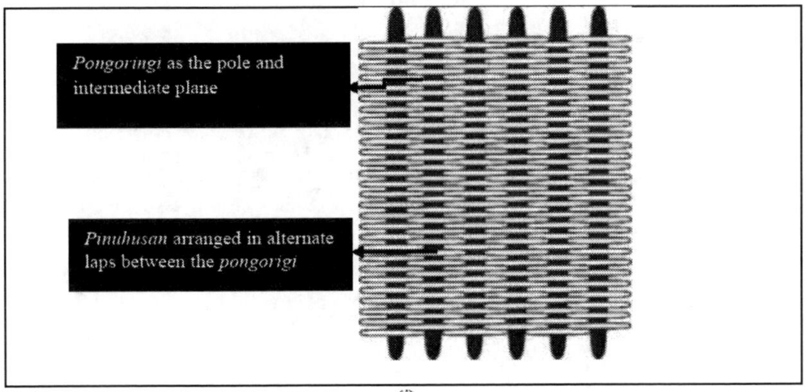

(i)

[3] Information from Konsimin binti Dalani (57 years old).
[4] Information from Sigim Binti Dalani (73 years old).

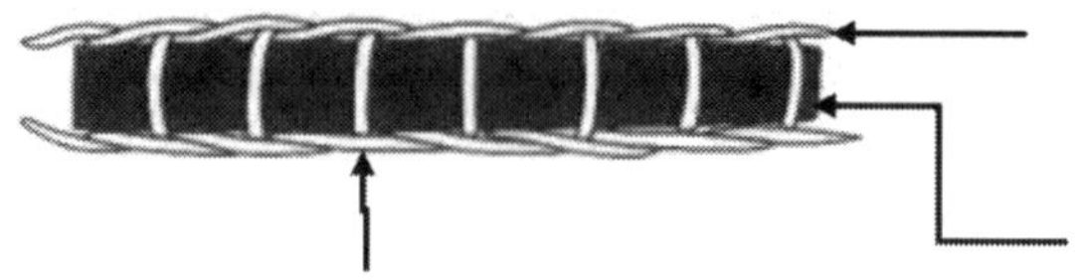

(ii)

Figure 3: ***Kinulung weaving technique***

Figure 3 (i) shows the result of *kinulung* woven in a square arrangement. When the weaving is completed, the weaver will gently push and manipulate it into the desired shape. This process using palm manipulation is referred to as *mongulung*. Figure 3(ii) shows the binding formed on the *dulung* or *kinulung* frame. This binding, from the finely sliced skin and soft inner part of the rattan, uses the interspersing technique whereby the twine is woven inside and outside of the frame structure using either the *kinagasan* (knot tying technique) or *binolilit* (winding between the softer inner rattan part on the *kinulung* frame) form of attachment.

3 Value and role of *kinulung* in the Dusun Tindal community

The *kinulung* craft of the Dusun Tindal nowadays are used more for decorative or commercial purposes only, and no longer serve as daily features in their kitchen. *Kinulung* could be found in the weekly Sunday market of Kota Belud and sold more often by handicraft wholesalers. The cultural element and heritage value of the *kinulung* has potential in making the craft a lucrative income generator which could sustain both weavers and the tradition alike. Dusun Tindal *kinulung* weavers would usually sell their products to individual buyers and wholesalers with the price range between RM10 to RM15 (for the former) and RM5 to RM8 each (for the latter).

To ensure the sustainability of the Dusun Tindal *kinulung*, several steps could be implemented: (1) disseminate knowledge of local terms and concepts in the weaving work, and (2) ongoing craftsmanship and documentation of aspects of design, technique and manufacturing material. The perseverance and patience of craftsmen reflected in the various stages of *kinulung* production is indicative of the innate creativity of the Dusun Tindal community especially among the womenfolk who have been integral as a source of thought and skill in the production of traditional bamboo-based crafts for generations.

4 Conclusion

Modern life in the era of rapid technological and digital advancement directly erodes the traditional and cultural values of a society. Modernisation has affected the art of *kinulung* as an integral heritage and cultural element in the Dusun Tindal community. Traditional handicraft is surely threatened with extinction if no preservation measures are put in place. The study and documentation on the design and production process of the Dusun Tindal *kinulung* showcases the local knowledge or expertise of the community and provides a viable channel for the dissemination of this legacy and skill for the younger generation.

References

1. Ismail Ibrahim. (2007). *Warisan Motif dan Corak Etnik Sabah*. Sabah: Penerbit UMS
2. Konsimin binti Dalani (57 years old), craft maker with weaving techniques. Kampung Tuguson Kota Belud. Interveiewd on 6 September 2016.
3. Siti Zainon Ismail. (1986). *Rekabentuk Kaftangan Melayu Tradisi*. Kuala Lumpur: Dewan Bahasa dan Pustaka Kementerian Pelajaran Malaysia.
4. Sigim Binti Dalani (73 years old), *kinulung* craft maker, Kampung Bukid. Interviewed on 10 October 2016.
5. Yahya Ismail. (1998). *The Cultural Heritage of Sabah*. Selanggor: Dinamika Edition.

Beaded shoes: the culture of Baba Nyonya

Noor Azam Azmi[1] , *Amirul* Nizam[2], *Diana* Mohamad[3] & *Badaruddin* Mohamed [4]
[1234] School of Housing, Building and Planning, Universiti Sains Malaysia

Abstract. *Despite the rapid changes in shoes' arts and designs due to globalization of cultures and customs, beaded shoes of Baba and Nyonya (else addressed as Baba Nyonya) remain in demand locally and globally. Baba Nyonya is one of the unique tribe or community in Malaysia because of their practiced lifestyle that harmonically combine both Malay and Chinese cultures. This paper descriptively discussed the basic procedure of making beaded shoes which include the philosophy and variations that still exist nowadays, in addition to discussing the technique applied when making beaded shoes. The paper also discussed about how the heritage of arts is still being applied by new generations these days where this is in parallel with the old quotes 'the time might not demolish the heritage arts'. The method employed by the researcher include revisiting literature regarding to the beaded shoes introduced by Baba Nyonya community. To earn a depth and detail information regarding to the beaded shoes, an interview was held with the informant which is one of the motivator or advisor that still practices the heritage arts of Baba Nyonya beaded shoes. The responsibility of continuing the culture is deemed vital and important so the new generation will appreciate and practice the culture practiced by Baba Nyonya.*

1 Introduction

Originating from China, the Baba Nyonya ethnic group or else known as the Straits-born Chinese are the descendants of the Chinese merchants who came to the Malay Archipelago during the 15th and 16th centuries. As these merchants settled and began families through mix-marriage, the Malay archipelago has become a melting pot for the uniquely blended community. In Malaysia, this exclusive and eminent ethnic can be found in two states namely Malacca and Penang Island (Worden, 2003), where the reside area was chosen due to commerce purpose. The Baba Nyonya, which practicing the distinctive lifestyle that combines both Malay and Chinese culture, is mostly living in Melaka (Lee, 2008). Nowadays, most Baba Nyonya exercised beliefs systems including Taoism, Confucianism, Buddhism, Christianity and Islam. The uniqueness of Chinese festivals celebrated by Baba Nyonya lies in the fact that Baba and Nyonya adapted Portuguese, Dutch, British, Malay and Indonesian influences in the customs. Other than culture, the individuality of Baba Nyonya could be observed in their traditional food, lifestyle and costume. Beaded shoes are one of it. This study is important because there are so many state treasures that are lost. Without any encouragement and effort to understand the culture of the arts of a nation has led the younger generation to move away from it.

2 Methodology

Information was collected through literature review pertaining to Baba Nyonya and beaded shoes, in addition to an interview session with Madam Lim Siew Seng in order to gain a authentic pictures and information regarding the Baba Nyonya's beaded shoes (see Figure 1). Information gained from the informant are applicable for use in improving the research (Marshall,1996). May Lim Siew Seng (aged 70 years old) is an active Beads-cobbler and it was learned that she has already engaged in making beaded shoes for almost 40 years, in addition she is also earnestly training the trainers who aim to be beaded shoes maker.

Figure 1. Informant, Madam May Lim Siew Seng

3 Literature Review

According to the history , the existence of Straits Chinese or Baba Nyonya ethnicity could be principally dated back to the 15th century (Khoo,1996). The history started when the close relationship with China was recognized when the Emperor of China sent his princess (Princess Hang Li Po) to Malacca as a token of gratitude to the Sultan of Malacca. Princess Hang Li Po and her servants decided Bukit Cina as their settlement area and eventually grew into Straits-born Chinese community. This Straits-born Chinee community known as Peranakans (else known as Baba Nyonya) is a result of mix-marriage between the Chinese and the Malay people. Interestingly, Baba Nyonya relates themselves with the modernism while still retaining the Chinese culture and tradition. To some extent the Chinese culture and tradition are flavoured mainly with the Malay culture and tradition elements and other religions' culture and tradition elements generally (Ahmad Az Mohd Mokhtar, 2012; Han, 2017). As a result of adopting, adapting and practicing various cultures, members of the Baba Nyonya community have mastered three languages: Chinese, Malay and British (Ahmad Az Mohd Mokhtar, 2012).

Currently, the population size of Baba Nyonya living in Melaka is only 600 person; and the population is getting smaller in size. Compunding this alarming fact is the realization that Baba Nyonya unique and eminent traditions and practices are being ignored by the new generations (Utusan Online, 2008). The unique characteristics of this cultures include (1KLIK, 2008; PYZAM, 2010):

1. Descended from China but were loyal to Britain.
2. Speak in Baba dialect where the dialect is Hokkien mixed with Malay languages.
3. Eat food with hand and all the recipes follow the Malys cokking style like using chili, belacan (blended small prawns), galangal acid and lemongrass.
4. The Nyonya (ladies) wear 'kebaya' and 'kain sarung'.
5. Houses are built following the Chinese traditional architecture style.
6. The Baba Nyonya wedding ceremony combines Chinese costumes (Figure 3) and Malays' traditional music (which is sung in Malay).
7. Cultural performance include Lion Dance, Fu Kien Opera and the Dondang Sayang Baba Nyonya.

Other than the aforementioned culture, the uniqueness of Baba Nyonya culture can be seen through the arts of embroidery, kebaya Nyonya , beaded shoes and crafts . Beaded shoes or else known as 'Kasut Manik' are handmade slippers with distinctive pattern created using embroided beads, and the shoes are usually made by the highly skilled artisan. However this highly skilled art are only limitedly inherited by the community in Baba Nyonya (Malaysia Aktif, 2016). The arts of making beaded shoes normally pass to the next generations in the family through family business (Utusan Online, 2008) and learning through observation (Sinar Harian, 2014).

The creation of beaded shoes (either using high quality fine beads or high quality rough beads) normally apply the flower motifs from Europe and high level of patience are needed during the production of the beaded shoes (Ahmad Az Mohd Mokhtar, 2012). In his blog entitled 'Transcrip of Kasut Manik Baba Nyonya', Milo Yahya (2012) stated that 'beads used on the shoes are made and imported from Europe, however, due to the high cost, later the beads used are imported from Japan. Despite the low cost, the Japan beads is of better quality and more oriental'. Baba Nyonya traditional beaded shoes motives consist of flora and fauna, and what makes these beaded shoes invaluable, is the fact that the motifs which has been used for a particular pair will not be used for other beaded shoes. The same motifs will be created only when it is requested by the same buyer.

According to WaiHaik Shoemaker which operates–in Melaka, beaded shoes is inhigh demand during the festives seasons. Other than the locals, these beaded shoes have successfully attracted buyers from Japan, Singapore and Germany. The process of making beaded shoes is as follows (Utusan Online, 22 Mei 2010 Accessed on 9 January 2017).
Requested motif is first drawn on the locating device with 1.5ft x 1.5ft measurement.

1. The beads are then sewed on the motif. Beaded shoes made with high quality fine beads will take approximately three to four months to complete (Grade 1 shoes will incur a cost upwards from RM1000 and Grade 2 will range between RM600 to RM900). Beaded shoes made with high quality rough beads will take approximately two months for completion and the prize ranged between from RM300 to RM450.

2. The completed sewn motifs will be cut accordingly and then, will be stamped on the footwear.

4 Informant Information

Beaded shoes are known as 'kasut manik bermuka' or 'face beaded shoes'. Usually and traditionally, the Baba Nyonya will wear beaded shoes when attending formal occasion (for example wedding ceremony) and when visiting family and friends during festive season. Given the nature of occasion of which beaded shoes are worn, the beaded shoes are extraordinarily designed and beautifully decorated. The quality of handsewn/handmade beaded shoes are of better quality compares to machine-sewn beaded shoes. Of importance, the quality of handsewn/handmade beaded shoes is instinctively controlled given the limitation of possibly created designs within a certain period of time. Further, any design or motives are limited to two or three pairs of beaded shoes. Hence, the authenticity value is preserved and maintained.

4.1 Philosophy of beaded shoes

Flora and fauna motifs are highly important in the creation of Baba Nyonya beaded shoes. Rose, natural flower and Teluki flowers are the few types of flower motifs used for the design (figure 2); while motives for fauna include birds, lion, dragon and chicken. According to the tradition and as one of the philosophy, the Nyonya (ladies) must wear matching color of beaded shoes and costume.

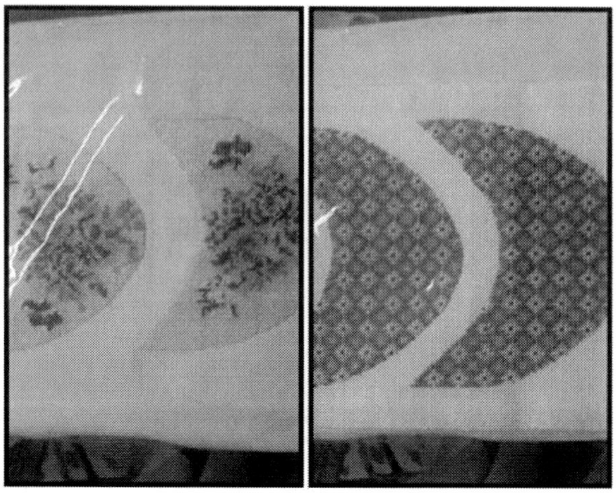

Figure 2. Beaded shoes motifs

4.2 Variation of beaded shoes today

To date, despite the fact that beaded shoes are expensive, they are still much in demand due to the authentic design as well as the difficulty of the making process . According to the informant, there are two different variation of beaded shoes, namely the traditional beaded shoes and the modern beaded shoes. Traditional beaded shoes are shoes that still maintain the flora and fauna motif. The material (beads) are imported from Europe because the quality is better and long lasting. Besides the quality aspect, the utilization of European beads emphasizes the distinction between social classes. Hence, indirectly, this indicates subject-positioning. Given these qualities, the traditional beaded shoes are sold at a notably higher price compared to the modern beaded shoes. The usage of traditional beaded shoes are limited to certain occasion or event for example wedding ceremony.

On the other hand, the beads used for the modern beaded shoes are imported from Japan and China. It is said that the beads from Japan are of better quality and the designs are more oriental-based when compared to the beads from China, which are slightly inexpensive. Therefore, modern beaded shoes made using the China beads are sold at a slightly cheaper price compared to modern beaded shoes made using the Japanese beads. In comparison to traditional beaded shoes, the designs for modern beaded shoes are more contemporary. What is meant by contemporary in terms of materials and pattern of use. The design is more modern with a more abstract and geometric pattern. Besides designed with heels, the modern beaded shoes also use pearls and crystal for decoration purpose. The reason for attaching pearls and crystal is to enhance the beads appearance (figure 3).

Figure 3. Modern beaded shoes

4.3 Technique applied in making beaded shoes

The following technique/procedure is applied when making both traditional and modern beaded shoes. As mentioned by the informant, the technique/procedure will take approximately 3 to 52 weeks to be completed, depending on the design (traditional or modern) and the materials used. This technique/procedure consists of 5 stages, namely:

Stage 1: Preparation of frame and silk cotton
The first step involves the preparation of the location device which measured half feet for both length and width. After the silk cotton is securely attached to the location device, the process of sketching the design will take place.

Stage 2: Sewing the separator or border between beads
Beads are sewed only on the separator/border sketch. This step needs to be completed before moving to the next stage.

Stage 3: Inner beads sewing
At this stage, the thread of complete sewn beads which located at the front of silk cotton will be ended and neatly tied at the back of the cotton. More importantly, this stage tests the credibility of the beaded shoes maker in producing an impressively attractive design. The maker, other than responsible of producing a high quality product, is also challenged in constructing a design that is beyond the expectation of the buyer. This is to maintain the authenticity value of the beaded shoes.

Stage 4: Preparation for footwear base
The footwear base, which normally made from 'Fomika Soo' type leather, will be drawn exactly following the customer foot size. The base will be cut in accordance to the sketches.

Stage 5: Patching the sewing
This stage requires the beaded shoes maker to patch the readily sewn beads to the footwear base. The patching process is done using glu (preferably strong) and after the completion of this process, a specific hammer will be used to neatly levelled the footwear base.

5 Knowledge transfer

The informant stated the reason why she is still actively engaging in this industry:~~is~~ because she wants to share the precious art of making beaded shoes to new generations who appreciate~~s~~ and understand~~s~~ the significance of beaded shoes. Classes and workshops organized by the informant have attracted participation from young locals and foreigners. Rather than instructing the trainees to follow the original ideas/designs, the informant has encouraged the trainees to further improvise the existing designs without compromising the meaning of Baba Nyonya culture and customs. Given the fact that the production of beaded shoes involves such a tedious process, it is undeniable that patience and industry are two of many other attributes needed to survive in this industry. Besides classes and workshops, the informant is also involved in the 'Program Pembelajaran Pertukangan Seni Tradisional', which is organized by the Penang Apprenticeship Programme for Artisans (PAPA) (Figure 4). PAPA's mission is ensuring the transfer of the traditional arts knowledge to the new generations. In addition to that, PAPA also invited the informant on overseas trips for the purpose of introducing the Baba Nyonya beaded shoes along with other products created by the artisans.

Figure 4. PAPA ;Skills and knowledge transfer program

6 Conclusion

The traditional procedure of beaded shoes making should be kept alive as this is significant in ensuring the sustainability of a certain culture in the rapidly transforming modern world. It has been proven that handsewn/handmade beaded shoes are of better quality, long lasting and display origin characteristic. Even though through modernization a lot of high technology equipments have been introduced, these equipments are still inferior to the human touch in terms of highlighting the purity and the authenticity of a product. Even the world value has changed, and the new generations are offered with ranges of products to choose from; sustainability of traditions and customs has to assured. This will help in preserving ones' identity and perception toward arts and culture. Therefore, it is important for the local knowledge of old generation to instill, to teach and to remind the new generations about the importance and the significance of beaded shoes culture and arts. Traditional arts and knowledge transfer program should be paid attention by the appointed organizations such as Penang Island, Penang Apprenticeship Programme for Artisans (PAPA).

Aknowledgement

The authors would like to extend the aprreciation to Kementerian Pengajian Tinggi Malaysia for the funding of Geran Penyelidikan Jangka Panjang (LRGS), (Kearifan Tempatan), account number 203/PTS/6727002 entitled Contemporizing Trandisiplinary Knowledge Transfer for the assurance of the research and the article. The authors also would like to thank the informant, Madam May Lim Siew Seng, for her great cooperation and valuable information.

References

1. 1KLIK (2008). Keunikan masyarakat Baba dan Nyonya. Diakses pada 9 Januari 2017 dari http://pmr.penerangan.gov.my/index.php/budaya/8718-keunikan-masyarakat-baba-dan-nyonya.html
2. Ahmad Az Mohd Mokhtar (2012). Kajian mengenai etnik Baba dan Nyonya dan hubungannya dengan rumpun Bahasa Melayu. Diakses pada 9 Januari 2017 dari http://sayauncleaz.blogspot.my/2012/11/etnik-baba-dan-nyonya.html

3. CRIonline (2016). Pameran kraf tangan Baba dan Nyonya di Singapura. Diakses pada 9 Januari 2017 dari http://malay.cri.cn/901/2016/06/24/241s163289.htm
4. Han (2017). Asal usul Baba dan Nyonya. Diakses pada 9 Januari 2017 dari https://www.academia.edu/8526027/Asal_Usul_Baba_Dan_Nyonya?auto=download
5. HardwareZone Malaysia (2016). 59 things that makes us Malaysians. Diakses pada 9 Januari 2017 dari http://www.hardwarezone.com.my/feature-hardwaremerdeka-59-things-makes-us-malaysians/hardwaremerdeka-59-things-makes-us-malaysians-habits-festivals-and-ethnic-costumes
6. Khoo, J. E. (1996). *The Straits Chinese: a cultural history*. Knickerbocker Press.
7. Lee, S. K. (2008). The Peranakan Baba Nyonya Culture: Resurgence or Disappearance?. *SARI: Jurnal Alam dan Tamadun Melayu*, *26*, 161-170.
8. Malaysia Aktif (2016). Usahawan tunggal Melayu Kasut Manik Baba Nyonya. Diakses pada 9 Januari 2017 dari http://malaysiaaktif.my/malaysiaaktif2/?p=20870
9. Milo Yahya (2012). Transcript of Kasut Manik Baba Nyonya. Diakses pada 9 Januari 2017 dari https://prezi.com/dsu0gcjp8hax/kasut-manik-baba-nyonya/
10. Marshall, M. N. (1996). The key informant technique. *Family practice*, *13*(1), 92-97.
11. Nazrulhad Hashim (2011). Festival antarabangsa Baba dan Nyonya. Diakses pada 9 Januari 2017 dari https://nazrulhad.wordpress.com/2011/12/20/festival-antarabangsa-baba-dan-nyonya/
12. Noraziah Hanafi (2012). Keunikan rumah Baba Nyonya di George Town serta budayanya. Diakses pada 9 Januari 2017 dari http://shegroup44.blogspot.my/2012/11/keunikkan-rekabentuk-rumah-baba-nyonya.html
13. PYZAM (2010). Adat dan budaya etnik di Malaysia. Diakses pada 9 Januari 2017 dari http://bmpengurusan2010.blogspot.my/2010/09/baba-dan-nyonya.html
14. Sinar Harian (2014). Kasut manik Umbai Melaka. Diakses pada 9 Januari 2017 dari http://www.sinarharian.com.my/rencana/kasut-manik-umbai-melaka-1.254903
15. Transez (2016). Peranakan. Diakses pada 9 Januari 2017 dari https://transez.wordpress.com/2012/07/22/peranakan/
16. Utusan Online (2008). Keunikan budaya Baba Nyonya. Diakses pada 9 Januari 2017 dari http://ww1.utusan.com.my/utusan/info.asp?y=2008&dt=0327&sec=Selatan&pg=ws_01.htm
17. Utusan Online (2010). Keunikan kasut bermanik. Diakses pada 9 Januari 2017 dari http://ww1.utusan.com.my/utusan/info.asp?y=2010&dt=0522&pub=Utusan_Malaysia&sec=Selatan&pg=ws_01.htm
18. Worden, N. (2003). National identity and heritage tourism in Melaka. *Indonesia and the Malay world*, *31*(89), 31-43.

Interview

May Lim Siew Seng, 66 Lebuh Acheh, Pulau Pinang, 6 November 2016

Malay Wisdom in the Motifs of Bamboo Shoots

Safial Aqbar Zakaria[1] *Ramlah* Latif[2], *Azizi Bahauddin*[3]

[1]Department of Interior Architecture, School of Housing, Building and Planning, Universiti Sains Malaysia, 11800 Minden, Pulau Pinang, Malaysia

[2] Department of Interior Architecture, School of Housing, Building and Planning, Universiti Sains Malaysia, 11800 Minden, Pulau Pinang, Malaysia

[3] Department of Interior Architecture, School of Housing, Building and Planning, Universiti Sains Malaysia, 11800 Minden, Pulau Pinang, Malaysia

Abstract. The study aims to reveal Malay's philosophical wisdom that is implied in Malay culture. Weaving is a part of Malay culture that is rich in historical and cultural value. Samples of the study are the *ikat limar* woven cloth, while the focus of the study is the bamboo shoots' motif. Bamboo shoots are one of the motifs synonymous with the *ikat limar* weaving. The bamboo shoots' motifs are located on the head of the sarong and on the edge of the shawl. Bamboo shoots in a tapered triangular shape is an implementation of the shape of the bamboo stems that appear from the culms. Bamboos are such a plant that brings a lot of benefits to the lives of the Malays from the shoots to the mature canes. The Malays often connect the bamboos with experiences, adulthood, beliefs, as well as pure values that shape the Malay ethos. Bamboo shoots are seen to own a broad sense of cultural philosophies that bring about the existence of sayings, expressions, poem, and figurative speeches. The objectives of the study are to explore the values of Malay wisdom in the motifs of bamboo shoots aside from introducing *limar* cloth as one of the heritages of Malay wisdom. The methods of the study are interviewing of the aesthetes, experts in textile field as well as those who are related to the textile and cultural field. The qualitative analysis will be used to combine the descriptions, analyses, and interpretations. Based on the outcome of the study, it will become a cultural document and source of reference for the young generations. This study also manages to attract outsiders to recognize and understand the Malay culture as well as getting to know the Malay nature realm.

1 Introduction

Bamboo shoots are one of the motifs that is synonymous within the realm of Malay's visual art. In the design or architecture of Malays' homes, for instance, the design of bamboo shoots is implemented upon the house' ridges or ridge caps. On the weaving of *limar* cloth and other Malay textiles, bamboo shoots' motifs are usually located on the head part of the sarong and on the both of shawl. The motifs of bamboo shoots are carefully displayed in various organized placement, which illustrates the creativity and competency of the weaver during the *ikat benang* process to produce such motifs. Such creativity and competency have managed to bring in aesthetical value to the *limar* cloth. This matter is conveyed through Abdul Hamid (1995), the beauty exists within a lot of objects and the visible objects' beauty can be seen through objects that can be seen. In reference to *limar* cloth, the object is not only seen upon the design or structure, it also includes the motifs.

The shoots of bamboo shoots are produced through the deep appreciation of the Malays towards the surroundings. It is also said by Ismail (1992) the Malays view some things not only focusing on the material side, they also have a deeper view on the world as seen by them. The statement explains how the Malays see things through physical view and spiritual view. Physical views refer to the tangible or extrinsic objects, while the spiritual views are the instrinsic objects (intangible), Based on this study, *limar* cloth and bamboo shoots are intangible objects. The meanings, values, or philosophies are the intrinsic values of a *limar* cloth.

The relationship of the Malays with nature is so close-knitted. This relationship is explained by how the Malays use natural resources in their daily lives. It demonstrates the existence of wisdom in interpreting the implied nature into work. The nature is a part of the Malays and how it is created is conveyed as a teacher. Through the nature and behaviour, the Malays have managed to produce figurative speeches, sayings, poetry, poems, idioms, and many more. All of this is of important platforms of the Malays in expressing their thoughts. As clarified by Deraman (2000), the Malays have long learned from the nature until such saying is born:

Baik-baik membelah buluh, (Be careful when splitting bamboo)
Dalam buluh ada ulat (There are grubs in bamboo)
Baik-baik mengkaji tubuh (Be careful observing the body)
Dalam tubuh ada sifat (The are characters in the body)
Baik-baik membuang ulat (Be careful when removing the grubs)
Dalam ulat ada sengat (There are stings in the grubs)
Baik-baik mengkaji sifat (Be careful when studying characters)
Dalam sifat ada zat (There are souls in characters)

The saying goes on how the wisdom of the Malays in selecting metaphors by using natural elements which illustrates how detailed they are in observing and grasping meanings. The saying is inclined towards the *tasawuf* teaching that touches on the spiritual and nature that are unbreakable or inseparable. The arrival of Islam in Malay world has given a new, better perspective to the Malay culture. Islam has shaped the perspectives of the Malays upon a better life by prioritizing knowledge as the core of life in this world and the Hereafter. The Malays are taught to see the world as a creation beneficial in life.

Wisdom of the Malays in "bamboo shoots" is a study to explore and interpret the wisdom of the Malays in relation to bamboo shoots motif. The motifs exist in the variety of objects created by the Malays own a broad sense of meaning and philosophy. Limar cloth is a material or visible object that can be seen and touched, which is fabricated by the Malays through ikat tenun technique.

1.1 Objectives of the Studies

The objectives of this study are to discern the wisdom of the Malays in the bamboo shoots motif, and later to comprehend the Malay's values conveyed through the bamboo shoots motifs. Next is to introduce *limar* cloth as one of the heritage of Malay wisdom.

1.2 Methods of the Study

This study also refers to previous studies through literature sources as supporting research findings. Conducting interviews is one of the methods used by interviewing the intellects and individuals involved directly in the culture field and the field of *limar* cloth weaving. Literature study, interviews and observations were made to support the findings of the study. Library studies refer to previous studies including viewing the source of sayings, expressions, poem and figurative speeches which is the mind of the Malays. Interviews were conducted through open-ended methods. Attention was given to the question of Malay culture, Malay cosmology in the motifs of the limar cloth weaving as part of Malay culture.
An interview was made with Mohamed Najib Ahmad Dawa, an experienced and expansive figure in traditional art, symbols and symbolism in Malay culture, architecture and modern art. Interviews were also made with individuals in Malay woven fields. The two leaders who have been interviewed were Norwani Md. Nawawi, songket experts and weaving specialists. Besides that, an interviewed was made with Manang Wan Awang, the third generation in Malay weaving. These figures are very close to the Malay fine art and have answered the questions of the Malay philosophy in the bamboo shoots as the subjects studied. Interviews were also made with the weavers community to scrutinize the views as well as the weavers' understanding of the chosen motifs. An interview session was held at the National Visual Art Center, Universiti Teknologi Mara (UiTM) and in the village of weaving in Terengganu. Observation was carried out simultaneously to see the nature and nature of the references of bamboo shoots and the activities of the local community reflecting on the meaning and philosophy of the study. Qualitative analysis combines observation and interpretation to the findings of this study.

 https://doi.org/10.1051/shsconf/20184502004

1.3 Definition of Keywords

In reference with the Kamus Dewan, wisdom gives a sense of courtesy, wisdom, cleverness and caution, intelligence explaining that a tactful person is a sensible person, knowledgeable referring to the meaning of being wise and cautious.

The wisdom of the Malays based on the study is of the courtes of the Malays as well as the values of wisdom, cleverness, caution, that are reflected through the motifs of the bamboo shoot. Motifs according to Bujang and Hamidon (2004) can defined as objects that form the basis of the element in design. Based on this study, motifs are decorative elements in *limar* cloth. Motifs are a part of the pattern and every motif has a meaning of its own. It becomes a symbol that provide its own value. The Malay culture refer to the ways of thinking, the manners of action received as a value by the Malays and the society which are reflected in the motifs of the bamboo shoots.

Limar cloth is a type of fabric woven by *ikat benang* method, colour-dyed and woven. Yarns are tied with a rope that can strap or block the color absorption of the strapped yarns. People in the older days used straps of banana heart, taken from banana leaves or banana stems. The yarn knot and the dyed colour form motifs on the fabric after being woven.

Bamboo shoots are located on the head section of the sarong and on both of the shawls.

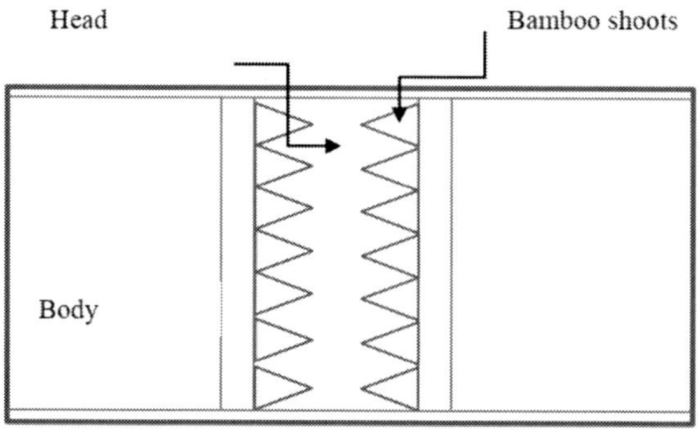

Fig. 1. Location of bamboo shoots on a sarong

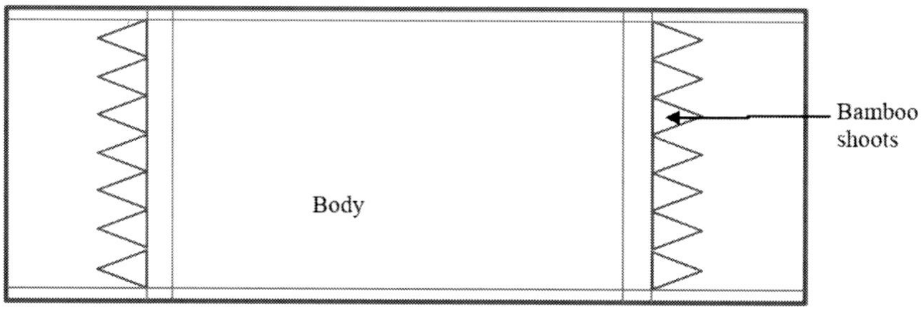

Fig. 2. Location of bamboo shoots on a sarong

Fig. 3. Bamboo Shoots motifs
Source: Terengganu State Muzeum

1.4 Types of Bamboo Shoots

A lot of variations and styles of bamboo shoots on *limar* fabric;

Fig. 4. *Lawi itik* bamboo shoot with *bunga susut sembunyi*
Source: Muzium Negeri Terengganu.

Fig. 5. *Kendung* Bamboo shoot
Source: Terengganu State Muzeum

Table 1. Styles and characteristics of the variations in bamboo shoots.
Source: Md. Nawawi (2016)

	STYLES	CHARACTERISTICS	SHAPES
1.	*Tunjuk Langit* Bamboo Shoot	The apex/pinnacle of a bamboo shoot. Tapered end looking upwards (space) or empty	
2.	*Kendung* Bamboo Shoot	Different types of bamboo shoots in alternating position	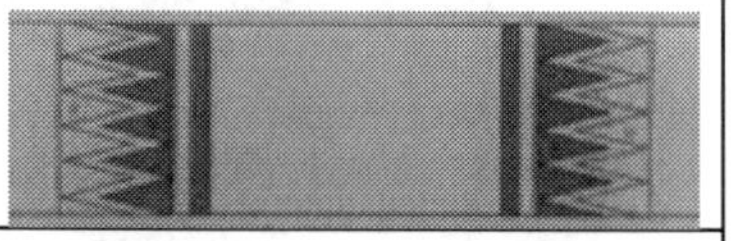
3.	*Gigi Bota* Bamboo Shoot	*Bota* is an ogre/giant. Bamboo shoot placed alternatingly, side by side	
4.	*Berjuang* Bamboo Shoot	*Berjuang* bamboo shoot refers to the meeting of the apexes/ tips	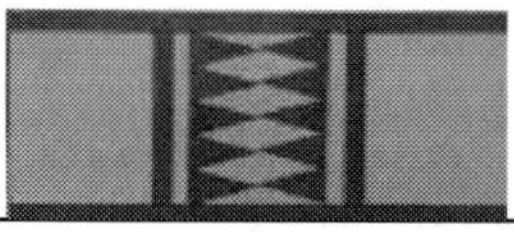
5.	*Siku Keluang* Bamboo Shoot	Zig-zag shape within bamboo shoot	

Table 2. Types of bamboo shoots.
Source: Md. Nawawi (2016)

BIL	TYPES OF BAMBOO SHOOTS
1.	*Lawi Ayam* Bamboo Shoot
2.	*Lawi Itik* Bamboo Shoot
3.	*Kayuhan* Bamboo Shoot

2 Discussion

The Malays often examine the nature in order to produce expressions and life philosophies while implied meanings are reflected through the Malay culture. The culture is connected to the wisdom of the Malays, either tangible materials such as weave, engraving, design, architecture, or the intangible, poetry, poem, sayings, and many others. The world refers to the natures of the fauna and flora as well as the nature of cosmo created for humankind. The world acts as the teacher or guide to the Malays which is quoted in the saying "*'the surrounding world as thy teacher*" that is explained by Muhammad Salleh (2006) the Malays think metaphorically by looking at themselves as the nature itself.

Efendy (2004) discusses about the tact and wisdom that are among of the nature of the Malays;

Arif menyimak kabar burung, (The wise will investigate)
bijak laku dalam bertindak, (Wise when taking action)

Next,

Yang arif menjemput tuah, (The wise invites luck)
Yang bijak menjemput maruah, (The clever invites honour)

The value of knowledge is highly regarded by the Malays. Tactful and wise people are fortunate and honourable. Tactful people is highly regarded in a society. This is due to the importances of the wisdom and knowledgeable people as references. That is why they are often elected as leaders in the society as such the saying goes:

Menjadi suluh dalam gelita, (He who enlightens the dark)
Menjadi penawar penolak bias, (Becomes the cure for venom)
Menjadi penenang dalam duka, (Who stays calm in sorrow)
Menjadi penyejuk dalam derita, (Who stays tranquil in agony)
Menjadi penyabar dalam duka. (Who in woe, stays unwearying)
(Efendy, 2004)

In reference with the bamboo shoots, wise people are defined from a lot of aspects based on experiences, observations, and interpretations of the Malays towards the behaviour and nature of the shoot as the foundation of the bamboo. Bamboos come from shoots. The shoots appear amongst the culms. The ability of the shoots to live and grow in the culms give a broad sense of meaning to the lives of the Malays. Bamboo shoots are linked to the perseverance and fighting spirit. Shoots are the soft stems that fight to stay alive amongst the mature canes in the culms. The upper-end tapered part symbolizes focus. This fits the explanation by Bahauddin (1999) which says motifs used are significant with the Malay culture and firmly states that the motifs are related to the concept of fertility, unity, and spirituality. On the other hand, bamboo shoots are related to the concept of nature, as explained the world is by layers. The cosmology of Malay-Hinduism for instance, perceive the cosmos to be consisted with levels and layers, which is the Trailokya (three worlds) concept: the upper realm, the middle realm, and the lower realm.

Zakaria (2015) further explains the layers of the realm refer to the nature of the human life. The world is divided into three symbolized by the bamboo shoot: Real world or universe (Alam Syahadah), Unseen World (Alam Mithal), Spiritual World (Alam Arwah) with the climax is the Divine Essence (Zat Allah). This depicts how wise is of the Malays in providing meaning and interpreting the symbol of bamboo shoots in relation to the greatness of God which become the creed's grasp of the Malays after reverting to Islam as the embraced and accepted religion. The world can be unravelled as follows:

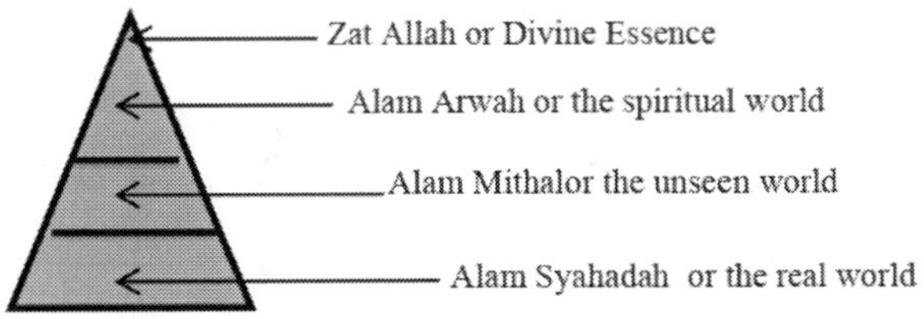

Symbol of Bamboo Shoot
Source: Zakaria (2015)

By referring to the layers above, Real World through the Malays' perspective is a visible world; skies, the earth as well as its substances rivers, hills, mountains inclusive of the Sun, the Moon, stars, oceans, plants and the world as a whole. Unseen World according to the Malays is a space that cannot be seen with naked eyes. Daud (2002) stated that the Malays divide Spiritual World into three, which are fairyland heaven, and hell. Meanwhile, Purgatory encompasses the spiritual world. Hereafter world, and the world before being born into the Earth. Divine Essence, the highest elevation of the Creator, the Almighty, remains as stated in the Quran, surah Al-Rahman (55);

> *"Segala yang dimukabumi akan binasa;yang kekal adalah zat Tuhanmu Yang mempunyai kebesaran dan kemuliaan."*
>
> *Surah al_Rahman (55), ayat 26-27.*

> *"Everything in this world will perish, And there will remain the Face of your Lord, Owner of Majesty and Honour"*
>
> *Ar-Rahman (55), verse 26-27*

Islam teaches the Malays on how to stay humble, recognizing that as a impermanent human and servant in this world, death is inevitable and everything that we hold dearly and proudly will be left behind. Islam seeping into a Malay perfects the modesty culture of the Malay. This is explained by Ahmad Jamal (1992), the most obvious cultural aspects of Malays' are the sense of modesty, staying humble and not boasting around about his or her expertise or wealth. The sense of humility is the essence of a traditional Malay society.

A bamboo shoot in a triangular shape according to Bahauddin (1999) explains there are four elements orderly arranged inside our bodies or humanself.

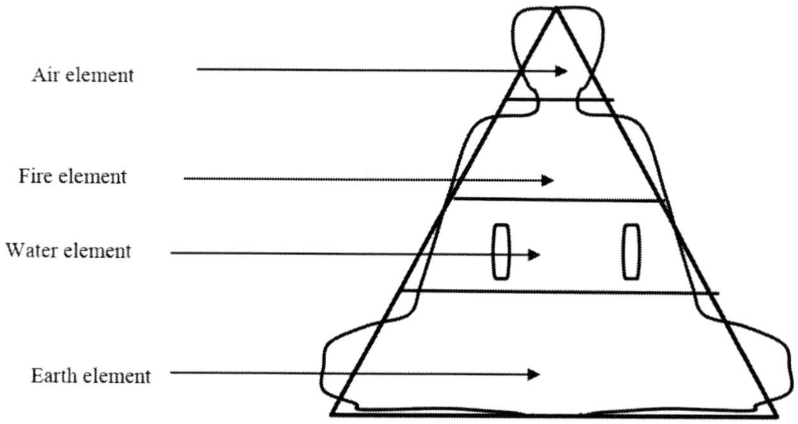

The contemporary Malaysian Arts: An Exploration of the songket motifs.
Source: Azizi Bahauddin (1999).

According to the views' of the Malays, the creation of the universe and humankind itself originated from four essentials or elements;earth, water, fire, and wind which inhibit the nature of cold, humid, hot, and dry. Seyyed Hosein Nasr (1993) states that the four elements and essentials develop ourself, which is known as microcosm.

On another words, bamboo shoots symbolize the process of personal development and self-development since young. Early childhood education plays an important role in shaping one's personal characteristics of becoming competent individuals to the society. According to Muhammed Najib Dawa via an interview, shoots personify a growing young child. Meanwhile, the bamboo symbolizes an adult. Adults are subjected as reference to the children. The Malays relate competent individuals whom are taught since young will be able to benefit the society as they grow older. For instance, the bamboos are made as walls, floors, roof bars, stairs as well as being used outside as fences, bridges, which often used by farmers and fishermen. At rivers, bamboos are designed as rafts, fish traps (*bubu*) or *tembilar* (similar to *bubu*), as well as being made into nesting box for the hens to lay and brood eggs. The importances of shoots maturing denote men who are beneficial and significant to the society like the saying goes:

Karam di laut jadi pelampung,
Panas di darat jadi penaung,
Basah di hilir jadi penudung,
Patah di hulu jadi penyambung.

Foundered in the sea, stayed afloat,
Blazed onshore, acted as a support,
Drenched downstream, perched with a hood
Broken upstream, befitted a bridge.
(Seyyed Hosein Nasr, 1993)

The wise Malays in examining bamboo shoots in various aspects of Malays' lives by looking through the nature, actions, and characters of bamboo shoots. Bamboo sprouts are bamboo stems grown amongst the bamboo culms. Bamboos can be preserved and cooked as appetizing dishes. Bamboo sprouts are symbolized as children whom are growing into adulthood.

To bend a bamboo, while it is still a shoot, this saying explains that bamboos are easily bent while they are still shoots. This shows the metaphorical ways of thinking of the Malays in establishing the nature as their teacher. In the Malay architecture for instance, bamboo shoots' motifs are visible upon the shape of the roof ridge of their houses. The triangular shape of the ridge is the bamboo shoot's shape that is related to the highest authority in this layered world, which is of the essence of Allah, as stated by Zakaria (2015) in the philosophy of bamboo shoots. In the everyday life of the Malays, the highest place is the roof of a house. It is tantamount to the Head, which is honourable for the Malays hence why it is intolerable to toy with their heads.

The ridge of a Malay house is an honourable and elevated place. The highest place demonstrates the place of the powerful and exalted essence. It is where the brain is located, which is used to think and decide upon rational actions. The Malays honour the matters related to humans' heads. Head accessories such as *kopiah* (*kopeah*) and *songkok* are often placed at the highest location that cannot be trampled upon or walked over. These behaviours displayed above show the perspectives of the Malays in regards with the importances of honouring knowledge as they regard highly the heads and matters relating to it.

3 Conclusion

The wisdom of the Malays in regards with bamboo shoots were reflected through personification, meanwhile, the struggle could be interpreted through the bamboo shoots' characters that emerged and became matured amongst the older canes. The strong philosophical belief of Malays in the essence of the Creator was explained through the nature hierarchy, starting from the real world to the essence of the Almighty at the highest elevation. This unfolded the humility of the Malays, which often realized that human life is fleeting, worldly gratifications will be left behind, the only that will remain is the Creator. The interpretations of values and philosophies in bamboo shoots' motifs had demonstrated the prudence of the Malays in scrutinizing and defining the world represented by bamboo shoots.

References

1. Abdul Hamid, Hamidah. (1995). *Pengantar Estetik.* Kuala Lumpur: Dewan Bahasa dan Pustaka. p, 5.
2. Ahmad Jamal, Syed. (1992). *Rupa dan Jiwa*. Kuala Lumpur: Dewan Bahasa dan Pustaka. p, 2.
3. Bahauddin, Azizi. (1999). *The contemporary Malaysian Arts: An Exploration of the songket motifs.* Tesis Doktor Falsafah (Ph.D), Sheffield Hallam University.
4. Borhan, Zainal Abidin. (2006). *Pantun Dalam Ungkapan Serta Pandangan Sementara Melayu dalam Pandangan Semester Melayu.* Kuala Lumpur: Dewan Bahasa Dan Pustaka. p, 247.
5. Bujang, Rahimah & Hamidon, Nor Azlin. (2004). *Kesenian Melayu.* Kuala Lumpur: Universiti Malaya. p, 96.
6. Daud, Harun. (2002). *Pemikiran Melayu tentang Alam dan Hakikat Diri* in Syed Muhammad Dawilah al-Edrus, *Pemikiran Melayu tentang Alam dan Hakikat Diri. Kumpulan Kertas Kerja Kolokium Bahasa dan Pemikiran Melayu /Indonesia II.* Kuala Lumpur: Dewan Bahasa dan Pustaka. p, 115-137.
7. Deraman, Aziz. (2000). *Tamadun Melayu dan Pembinaan Bangsa Malaysia.* Kuala Lumpur: Dewan Bahasa dan Pustaka. p, 30.

8. Effendy, Tenas. (2000). *Pemimpin Dalam Ungkapan Melayu*. Kuala Lumpur: Dewan Bahasa Dan Pustaka. p, 8.
9. Effendy, Tenas. (2004).*Corak dan Ragi Tenun Melayu Riau*. Jogyakarta: AdiCita. p, 83, 87.
10. Ismail, Moktar. (1992). *Rumah Tradsisional Melayu Melaka*. Kuala Lumpur: Persatuan Muzium Malaysia. p, 49,51.
11. Ismail, Siti Zainon. (2006). *Pakaian Cara Melayu*. Bangi: Universiti Kebangsaan Malaysia. p, 235. *Kamus Dewan Edisi Ketiga*, (1996) Kuala Lumpur: Dewan Bahasa Dan Pustaka, p, 815.
12. Md. Nawawi, Norwani. (2016). *Ikat Limar The Ancient Malay Textile*.Kuala Lumpur: Dewan Bahasa dan Pustaka. p, 12.
13. Md. Nawawi, Norwani. (2017, April 28). *Motif-motif dalam Tenunan Kain Limar* (Interview).
14. Mohd Yusoff, Zulkifli. (2010) *Kamus Lengkap Al-Quran*.Kuala Lumpur: PTS Publishing House.
15. Mohamed Najib Dawa. (11 March 2016). *Falsafah dan simbol motif-motif seni tampak Tempatan* (Interview).
16. Muhammad Salleh. (2006). *Dalam Daun Ada Bicara:Falsafah Alam Pantun Melayu*. Kuala Lumpur: Dewan Bahasa & Pustaka. p, 1-34.
17. Seyyed Hossein Nasr (1993) *Pengenalan Doktorin Kosmologi Islam*. Kuala Lumpur: Dewan Bahasa Dan Pustaka. p, 167.
18. Wan Manang. (2016, June 12). *Falsafah Motif Kain Limar* (Interview).
19. Wan Daud, Wan Mohd Nor. (2005) *Kerangka Keilmuan Tamadun Melayu: Satu Cerminan Ketamadunan*. Kuala Lumpur :Dewan Bahasa Pustaka. p, 98-120.
20. Zakaria, Safial Aqbar. (2015). *The Interpretation of Light in Bamboo Shoot Malay Songket motif*. Thesis (Ph.D). Universiti Sains Malaysia.

Contribution of local knowledge towards urban agroforestry as a sustainable approach on climate change adaptation

Jasmin Irisha Jim Ilham[1], *Norizan* Esa[1*]
[1]School of Educational Studies, Universiti Sains Malaysia, 11800, Pulau Pinang, Malaysia

Abstract. A crucial role in climate change adaptation is the one that should be played by cities. These are major contributors to climate change as well as most vulnerable to the effects of climate change. As urban population in cities increases, so does the greenhouse gas emissions. Urbanisation and the growth of cities continues at great speed, and today the global urban population has increased nine fold since 100 years ago. Though presented with current challenges and threats such as environmental degradation, urban poverty and food security, urban density could create a better quality of life and a lower carbon footprint by innovative approach and more efficient infrastructure planning. Creative solutions are implemented as part of the sustainability component of an integrated, climate resilient city, which comprises of fundamental aspects such as food, water, energy and transport. Application of local knowledge on agroforestry can be utilized in an urban setting, and has positive impacts on a microclimate scale, such as reduces impact of higher rainfall, reduction of greenhouse gas emissions and reduces vulnerability on the urban poor. Urban agroforestry largely contributes to sustainable urban food production opportunities, and encompasses four main fields of practice, including agroecology, urban forestry, urban agriculture and permaculture. This paper examines the local knowledge of agroforestry and how the local knowledge can be transferred into practice in an urban setting, thus providing a sustainable approach towards climate change adaptation in cities.

1 Introduction

According to the United Nations Populations Division (2014), the population of cities is projected to increase to five billion by 2030, signaling the largest wave of urban growth in recorded history. This situation arises the issue of food and nutritional security as one of the biggest challenges in urban ecosystem. An urban ecosystem can be classified as a group of interacting species that involves humans and their local, non-biotic environment working together as a unit. It can be separated into one single ecosystem or comprised of several individual ecosystem elements such as neighborhoods, watersheds, parks and lakes.

Integration of agroforestry in an urban setting could enhance the sustainability of the area, as well as combat the climate threats that have been evidently occurring at present. Urban agroforestry can reduce the ecological footprint at the certain area by increasing the production of ecosystem services [1]. Agroforestry can be defined as an intensive land-base management practice that integrates trees and woody perennials with agricultural crops and/or livestock. This form of practice optimizes the beneficial interaction of biological, physical and ecological level while ensuring economic needs of the farmer and his community are met [2].

Unique inclusion of trees and woody perennials contributes significantly to the diversity of agroecosystems, which provides soil conservation services, mitigation of pollution at point source, preservation of habitat and sustainability of agronomic products including organic crops [3]. Besides that, agroforestry practices can also include riparian and forest buffers, multi-strata home gardens, windbreaks and temperate zone analog [4].

Complex agroforestry systems depends greatly on processes, be it social or ecological in its management. Based on formal sources, farmer-based system management is being utilized through plating schemes, levels of biodiversity and selection of species [5]. However, transfer of agricultural knowledge is also done via social interactions [6] and so its development recognizes the necessity of including local knowledge [7]. In this context, local knowledge, or is also known as indigenous knowledge, means the cumulative and complex forms of knowledge, know-how, representations and practices that are being utilized and developed by peoples who have interacted with the natural environment for a very long time through the ages [8].

2 Methodology

Literature searches of electronic journal databases were conducted in order to gather information on contribution of local knowledge towards urban agroforestry as a sustainable approach on climate change adaptation. The key words used were 'local knowledge', 'urban agroforestry', 'cities', 'climate change', 'climate change adaptation'. Publications on research related to local knowledge on urban agroforestry, including review articles and published in English are criteria used for inclusion in the search to improve efficiency

3 Result and Discussion

3.1 Urban Agroforestry

Agroforestry in an urban setting is still quite uncommon in Malaysia. The integration of woody species such as trees and shrubs together with crops, livestock, or both in a farming system is more familiar in the smallholders sector as opposed to general practice [9]. Since the 1990s, agroforestry was perceived as a form of integrated farming that forms a promising and growing sector in agriculture, but it has yet to become a major undertaking in Malaysia as there are many challenges that accompanies it such as lack of knowledge and budget constraints [10].

There is a rising trend in urban agroforestry practice in home-gardens. Urban gardens in the Pacific Islands present a variety of food trees and non-tree plants that provide supplementary and staple food, including areas like Kiribati and Nauru that is known to have no agricultural diversity [11]. Many of the home gardens surveyed in Papua New Guinea, Fiji, Tonga, Kiribati and Nauru suggested that important staple root crops and supplementary non-tree food plants such as amaranths, onions, peanuts, pineapple, cabbages, as well as a variety of legumes, spinaches, okra, tomatoes, eggplant, corn and condiments such as ginger, coriander and mint were present in large proportion [11].

Home-gardens can also be in multi-strata, which typically consists of multiple species of plants occupying various spatial levels. With this method, energy capture and productivity of the system can be maximized. The multi-strata home-garden is a traditional agroforestry land-use system that include multipurpose trees and shrubs, herbaceous perennials and agronomic crops in polycultures located in close proximity to individual homes [12]. These productive polyculture system is adopted by permaculture and urban agriculture practitioners in cities worldwide as templates for forest gardens or food forests.

Clark and Nicholas (2013) proposed the term Urban Food Forest (UFF) that refers to agricultural practices located in home gardens, community orchards and perennial urban agriculture [13]. As a form of urban agroforestry practice, urban food forestry presents urban edible landscapes that includes food bearing perennial woody species strategically grown in to improve resiliency and sustainability of urban communities via integration of design principles and agro-ecological as well as plant breeding sciences. This in turn provides an opportunity to increase urban landscape productivity.

A case study by Mohd. Salleh et. al. (2016) from the School of Industrial Technology at Universiti Sains Malaysia shows how plants are grown at rooftop gardens to produce local food for the urban community [14]. Data as of 2016 shows a variety of plants being grown on the roof top garden, which consist of 46 plant species identified. The plant types that are present include herbs (21 species), shrubs (12 species), trees (4 species), climbers (4 species), aquatics (3 species), vine (1 species) and fern (1 species). The plant selection is heavily influenced by traditional indigenous gardening knowledge, specifically the Malay community. This is proven by the functionality of plant species, which consist of ornamental, aromatic and shade purposes besides sustainable local food production [15].

3.2 Application of Local Knowledge from Rural to Urban Agroforestry

The process of urbanization is a threat to practitioners of local knowledge in Malaysia. Preservation of local knowledge is important as it is a comprehensive, dynamic system. Within this system, people are able to continue with their current practices yet at the same time adapt to difficulties encountered in their rapidly changing environment and economic situation that may be challenging to their existence [16].

Most local knowledge that are being practiced focuses on rural agroforestry. This is largely due to resources of man power and persistence to improve economic income in household. The economic status of farmers is also a major factor that contributes to their involvement in agroforestry systems. In Malaysia, an allocation of RM297 million and RM7000 per hectare are distributed to smallholder farmers by the Ministry of Plantation Industries and Commodities (MPIC) and Rubber Industry Smallholders Development Authority (RISDA) [17, 18]. This financial assistance encourages micro level farming system and helps improve the agricultural productivity.

The Sabal Agroforestry Project in Sarawak is an exemplary project that applies local knowledge with participants from six longhouses within the vicinity [19]. The project aims to reduce shifting cultivation inside the permanent forest estate by offering land, materials and incentives to motivate farmers to practice a more sedentary form of agriculture, at the same time increase income and decrease poverty level in rural areas. One of the agroforestry programmes that were implemented at Sabal Forest Reserve is intercropping cash crops with tree crops. This allows local people to plant fruit trees such as durian (*Durio zibethinus*) in between rows of existing cocoa plants. Other agroforestry programmes that were carried out include planting of fruit trees, coffee and pineapples at existing forest plantation (*A. mangium and Shores spp.*). The objective is to maximize land use and as a trial basis for compatibility for short-rotation agricultural crops such as coffee and fruit trees. Fish pond and beekeeping were also involved in the programme.

The concept of agroforestry in general is to increase the sustainability of agriculture, regardless of the geographical setting. The main types of local knowledge approach towards agroforestry include home-gardens, shifting cultivation, plantation crop combination and various simultaneous and sequential systems [20]. Out of these approaches, the most common one that is being practiced in an urban setting is home-garden. Young (1989) vouched that this form of agroforestry is highly productive, fully sustainable and very implementable [21]. One of the key advantages for this practice is the labour efficiency due to close proximity to the farmer's residence.

Local influences are evident in settings of home-gardens in Malaysia. The cross disciplinary of practice between agroforestry and ecosystem services such as providing habitat for sunbird, stingless bee and honey bees has made urban farming more unique and attractive. Traditional practices are applied not only on edible plants but also to cater to the co-existence of present biodiversity. This sort of setting can be considered as application of biomimicry technology as the ecosystem replicates several key features of local landscape vegetation that is practiced in traditional house compounds [22]. Greeneries that is formed via home-gardens are visually captivating and is considered as an added aesthetical value for urban landscapes.

The practice of local knowledge in an urban setting is constantly evolving as it needs to be highly adaptable to change. In some circumstances, practices are affected by technological interventions, which greatly enhance the production of crops in an urban as compared to rural setting. Although this improvement contributes towards efficiency, it is more likely that that local solutions in rural areas are less costly and risky as compared to urban areas, where external introductions are involved. Communications is key to ensure that traditional agro-ecological knowledge is being conveyed coherently in order to not only maintain good agroforestry practice, but also as a tool to develop productive participatory relationship with local community, thus engage socio-culturally with one another [23].

3.3 Urban Agroforestry: Strategy for Climate Change Adaptation

Climate change is recognized globally as a threat. The World Bank (2010) encouraged new and innovative solutions for adaptations to climate change through environmentally sustainable solutions that involves integrated components of cities as a form of disaster risk management plan [24]. An example identified by The United Nations is urban and per-urban agriculture and forestry (UPAF). A selection of literature review indicates that UPAF has contributed greatly to the urban in enriching their food security, improving the city's natural greenery and urban local climate [25].

The concern by local government institutions and municipal councils for sustainable urban and regional planning and development should encourage them to proactively coordinate and implement agriculture and forestry in efforts to enhance food security and increase city resilience. Urban agroforestry contributes positively towards climate change adaptation. For example, by applying agroforestry on steep slopes, it reduces the impacts of high rainfall by minimizing landslides. Implementation and maintenance of urban agroforestry provide much needed shade and improve evapotranspiration, thus reduce the heat island effect. This is a form of passive cooling approach to reduce high temperature in urban areas [26]. Besides that, other particles such as carbon dioxide and dust are captured.

Use of energy in urban areas and emission of greenhouse gases can be reduced via urban agroforestry practice. The local knowledge being implemented in the sustainable process of planting such as composting is a way to reuse organic waste that is productive. This will in turn decrease the amount of organic waste sent to landfills, thus effectively cause less emission of methane gases from landfills. With more compost being produced, the demand for fertilisers will decrease and in doing so, less energy is needed to produce artificial fertilisers. At the same time, the carbon footprint related to the transport of food, storage, cooling and packaging will be reduced with more fresh food available close to home in the city.

4 Conclusion

The practice of urban agroforestry in an urban setting is indeed an emerging concept that is still slowly being implemented in Malaysia. However, there is a strong local knowledge influence in agriculture and forestry in urban settings that deserves to be preserved so that the local knowledge can be continuously be applied in future generations to come. It is important that the culture is being shared among the local communities to empower and encourage more people to start their own home-gardens and planting within their home vicinity. It is evident that urban agroforestry is an important player in the adaptation to climate change process. Thus, in order to create better and livable cities, integration of urban agroforestry is definitely the way forward, complemented and driven by the local knowledge for a more sustainable future.

Acknowledgement

The authors express their gratitude to the Malaysian Ministry of Higher Education for awarding the Long-Term Research Grant (LRGS, 203/PTS/6727002), Universiti Sains Malaysia for the Research University Grant (1001/PGURU/816274) and also all involved in this study.

References

1. Mann, Urban Agroforestry: Connecting agroecology, permaculture, urban forestry and urban agriculture with agroforestry. Forest 8385, *Ecological Principles of Agroforestry*, University of Missouri, Columbia (2013)
2. Lassoie, J. P., L. E. Buck, and D. Current, The Development of Agroforestry as an Integrated Land Use Management Strategy Chapter 1 in Garrett, H.E., (ed.). 2009. North American Agroforestry: *An Integrated Science and Practice*, 2nd edition, American Society of Agronomy, 677 S. Segoe Rd., Madison, WI (2009)
3. Gold, M. A. and H.E. Garrett, Agroforestry Nomenclature, Concepts, and Practices. Chapter 3 in: Garrett, H.E., (ed.). 2009. *North American Agroforestry: An Integrated Science and Practice*, 2nd edition, American Society of Agronomy, 677 S. Segoe Rd., Madison, WI (2009)
4. USDA Agroforestry Strategic Framework, Fiscal Year 2011–2016 (2011)
5. Beer, J., Muschler, R., Kass, D., Somarriba, E., Shade management in coffee and cacao plantations. *Agroforestry Systems* **38**, 139– 164 (1998)
6. Conley, T. and Udry, C., Social learning through networks: the adoption of new agricultural technologies in Ghana. *American Journal of Agricultural Economics*, **83**, 668–673 (2001)
7. Isaac, M. E., Dawoe, E., Sieciechowicz, K., Assessing Local Knowledge Use in Agroforestry Management with Cognitive Maps. Environmental Management, **34**, 1321-1329 (2009)
8. UNESCO, Chapter 13: Enhancing Knowledge and Capacity. In United Nations Educational, Scientific and Cultural Organization (UNESCO), *Water a shared responsibility*. New York, USA: Berghahn Books & Paris, France: UNESCO (2006)
9. Jaafar, A. A., Kasiran, N., Muhammed, S., Wan Ismail, W. H., Agroforestry Practices in Malaysia – Integrating Plantation Crops with Timber Species (2014)

10. Nazip, S. and Suhaimi, M., *Agroforestry Practices in Malaysia: Factors that Influence its Implementation. Bureau of Research And Consultancy,* Universiti Teknologi MARA, Shah Alam, Selangor, Malaysia (1998)
11. Clarke, W.C. and Thaman, R.R., *Agroforestry in the Pacific Islands: System Sustainability.* United Nations University Press, Tokyo, Japan (1993)
12. Fernandes, E.C. and Nair, P.R., An evaluation of the structure and function of tropical homegardens. *Agricultural systems,* **21 (4),** 279-310 (1986)
13. Clark, K.H. and Nicholas, K.A., *Landscape Ecol* **28**, 1649 (2013)
14. Mohd. Salleh, A., Azman Lim, N.A., Azam, M.A., Abdul Hamid, M.H., Ibrahim, M.H., Esa, N., Understanding the Adoption of Malay Landscape in a Rooftop Garden Design. USM Press (2016)
15. Rostam, K., Ahmad, R., Hussain, M.Y., Urbanization and the Changing Habitat: The Case of the Malay Housing Compound in Malaysia. *Jurnal E-Bangi,* **1(1),** 1–14 (2006)
16. Richards, P., *Indigenous agricultural revolution: ecology and food production in West Africa.* Hutchinson, London (1985)
17. Sinar Harian, Risda sedia RM18 juta untuk pekebun getah, sawit (2012)
18. Utusan Borneo, RM297 juta bantu pekebun kecil di bawah NKEA Kelapa Sawit (2011)
19. Kendawang, J.J., The Development and Application of Agroforestry Technology for the Rehabilitation of Tropical Forests in Sarawak, Malaysia. *JIRCAS,* Working Report No. **60,** 9-16 (2009)
20. Long, A.J., and Nair, P.K., *Agroforestry system design for the temperate zones: Lessons from the tropics* 133-139. In: Garrett, H. (Ed) Proceedings for the Second Conference on Agroforestry in North America. University of Missouri, Columbia (1991)
21. Young, Agroforestry for soil conservation ICRAF Science and Practice of Agroforestry series BPCC. Wheatons Ltd, Exeter, UK (1989)
22. Oberndorfer, E., Lundholm, J., Bass, B., Coffman, R. R., Doshi, H., Dunnett, N., Green Roofs as Urban Ecosystems: Ecological Structures, Functions, and Services. *BioScience,* **57(10),** 823–833 (2007)
23. Joshi, L., Arevalo L., Algre J., Sinclair, F., *Local Ecological Knowledge in Natural Resource Managment.* Paper presented at the conference: Bridging scales and epistemologies, Alexandria, Egypt, 17-20 May 2004, 16 p. (2004)
24. World Bank, Cities and Climate Change: An Urgent Agenda. Washington DC (2010)
25. Dubbeling, M., Urban and peri-urban agriculture and forestry as a strategy for climate adaptation and mitigation. *Sustainable Development Knowledge Platform* (n.d.)
26. Kamarulzaman, N., Hashim, S. Z., Hashim, H., Saleh, A. A., Green Roof Concepts as Passive Cooling Approach in Tropical Climate – An Overview. *E3S Web of Conferences 3,* 01028 (2014)

Tropical Wetland Valuation: An influence of local knowledge in Malay traditional ecosystem for human well-being

Farah Mastura Rosli[2], *Anizah* Salleh[2], *Widad* Fadhlullah[3], *Mahamad Hakimi* Ibrahim[3], and *Norizan* Esa[1*]

[1]School of Educational Studies, Universiti Sains Malaysia, 11800, Pulau Pinang, Malaysia
[2]Faculty of Technical and Vocational, Universiti Pendidikan Sultan Idris, Tanjong Malim, Perak Darul Ridzuan 35900, Malaysia
[3]School of Industrial Technology, Universiti Sains Malaysia, 11800, Pulau Pinang, Malaysia

Abstract. Tropical wetlands are among the most productive ecosystems on earth which provide vital services and consist of various types of plant and animal communities that live in the water and on land. However, this golden area is one of the most undervalued ecosystems and the awareness on protecting the areas is still lacking among communities. The aim of this study was to review the wetland services in a tropical climate which functions to complete the ecosystem. In particular, this study will i) explore the values of wetland ecosystem towards human well-being in tropical culture and ii) to design a complete wetland ecosystem with respect to the local knowledge in a tropical climate. This study indicates that the new millennium of ecosystem services provided by tropical wetland has been disaggregated into provisioning of goods; regulating services, cultural development and supporting the process by considering the influence of Malay traditional wetland. Therefore, a complete tropical ecosystem is designed by imitating the natural wetland for societal well-being.

1 Introduction

Wetlands play a vital role in human societies, accommodating a habitat and source of food for various types of plants and animals, supplying clean and retaining water naturally. Wetlands also serve as reservoirs for flood abatement and perform as a carbon storage. Due to rapid development worldwide, more than half of the original wetlands in the world are degraded. In 2005, the area of wetland that covered the earth's surface is 9%. But in 2011 the total area has decreased to 6% [1, 2]. Human activities such as dredging, installation of the drainage system for agriculture, and extensive agricultural activities near the water body have caused various environmental problems. The loss of wetlands vegetation threatening fisheries, damaging water quality and triggering eutrophication problem for which ultimately leads towards hypoxia and a dead zone for wetlands.

According to the most widespread definition, wetlands are defined as lands transitional between terrestrial and aquatic ecosystems where the water table is usually at or near the surface or the land is covered by shallow water [3]. However, according to Ramsar Convention on Wetlands define wetlands as areas of marsh, fen, peatland or water, whether natural or artificial, permanent or temporary, with water that is static or flowing, fresh, brackish or salt, including areas of marine water the depth of which at low tide does not exceed six metres [4]. Through layman language, wetlands are shallow or sporadically flooded ecosystems such as swamps, bogs, marshes, and sedge meadows [5].

Wetlands are critically important ecosystems, providing significant social, economic and ecological benefits such as regulation of water quality and quantity, habitat for water flow, fish and amphibians, resources to meet human needs, recreation and tourism. Many commercially important fishes and shellfish such as crabs, prawns, catfish, oyster, tilapia fish, and shrimp call wetland as their home. Several types of wetlands are known to act as hydrological buffers. For example, floodplain wetlands store water when rivers over-top their banks, reducing flood risk downstream [6]. The value of these services may be considerable and often technical alternatives to regulate the quantity of flow are much more expensive. The maintenance of many, complex, biological processes involving soils, water, plants, animals, and microorganisms, is necessary to sustain these ecosystem services. The functioning of a wetland ecosystem gives rise to a wide diversity of species as they support important levels of global biological diversity, including over 10,000 species of fish, over 4,000 of amphibians, and numerous species of waterfowl [7,8].

The effort to conserve and restore wetland ecosystems led to the creation of laws, regulations, and plans to restore and protect wetlands around the world. In the United States, the protection of wetlands is clearly fallen under The Clean Water Act of 1972, which permits for dredging and filling activities is required and to monitor the standards of water quality. However, there is no comprehensive legislation that relates to wetland conservation as a whole [9]. However, the resolution to conserve the wetlands in Malaysia have been spearheaded by Malaysia's oldest and largest environmental conservation group such as Malaysian Nature Society (MNS) under Wetland Program, Flyway Campaign and Wetlands Education, Resource and Training Centre for Communities [10].

Accordingly, this paper reviews the importance of wetland for human well-being, wetland valuation for a sustainable ecosystem, and how people can appreciate wetland by creating a mini wetland. We also provide a method to create Malay wetland design by using our adapted plants and animals to complete an ecosystem in a wetland.

2 Wetland as an ecosystem services

2.1 Wetland services

Wetland ecosystems provide important services that benefit people for their well-being. The Millennium Ecosystem Assessment has proposed four classes of services provided or derived by wetland by considering benefits obtained by people [11]. These include (1) provisioning services such as water and food; (2) regulating services such as erosion regulation, hydrological flows, mitigation of climate change, and water purification and waste treatment; (3) supporting services such as distribution of biodiversity, soil formation and nutrient cycling; and (4) cultural services such as recreational, spiritual, religious, and education purposes.

The loss of wetland areas affecting the lost of the key ability of ecosystem services. One of the services of wetland is it provides clean water for domestic, industrial and agricultural use which is underlined in provisioning services. Other than that, wetlands also contribute a large amount of protein such as fish. However, human activities have led to reducing their quality services. For example, water quality degradation by human activities especially by agricultural activity has led to eutrophication, which is one of the common problems for the wetland [12]. This phenomenon occurred by an excessive amount of nutrients from the use of fertilizer in agricultural activities which will stimulate the growth of algae in the water bodies. The decaying process of algae demands oxygen from the water body which will reduce the oxygen concentrations and ultimately kill fishes and disrupts aquatic organisms. Without mitigation measures, it will lead hypoxia to occur, where it is occasionally toxic to humans.

Flood is one of the natural hazards occurs which highlight the importance in maintaining the ecological function of wetlands. By flooding phenomenon, the organic matter is able to be transported into wetlands naturally which could sustain other important services such as food supply to the aquatic organism and water regulation [13]. Lakes, large river, estuary, and floodplain are the good example to attenuate the potential of flooding. The diminishing area of wetlands increases the risk of floods occurring. For an example, soil erosion affected by deforestation near the wetland can increase in soil content in the river and ultimately the wetland become shallow. When heavy and long rainfall happens, the river cannot support the amount of water coming in. Thus, it causing flood disaster.

Human intervention near the wetland modifies the energy and materials flow in the bank. This modification can affect the stability of food web in the wetland [14]. Migration of invasive species and deletion of adapted species in the wetland may lead to the energy inefficiency. Accordingly, it impacts the productivity of higher trophic level of the food web and destabilizing food web characteristics in the wetland [15,16]. Ultimately, it threatened the biodiversity of the wetland ecosystem.

As for cultural service, wetland also acts as one of recreational and education area for communities. For example, Shah Alam Lake in Selangor has become a recreational park. Various activities can be done by locals such as jogging, cycling, and canoeing near the lake. Other than that, Kenyir and Temenggor lake also have become as tourism area. Additionally, many researchers from higher institution in Malaysia conducted their research in the wetland [17, 18].

2.2 Wetland valuation for sustainable ecosystem

There are two most important wetland ecosystem services affecting human well-being involve fish supply and water availability. In a tropical wetland, freshwater fish species have higher demand compared to marine species. For example in Cambodia, 60-80 % of protein source was obtained from rivers for rural communities [11]. Besides cooked fresh, there are many commercial fish products in the market that become basic needed in cooking such as fish oil, fish sauce, dried fish and processed food.

Other than that, the wetland is known as a major source of clean water supply to the community. Clean water supply is important for the operation of a nation. In Malaysia, it is estimated that 76% of water was used for agriculture, 11% for municipal water supply and 13% for industries and only less than 1% of available water resources is used for drinking water supply [19]. Malaysia is drained by a dense network of rivers and stream which supply major clean water source. Construction of dam to collect and restrict water from tributaries and their source to the river as a method to collect clean water supply. Almost 56 dams were constructed around Malaysia such as Temenggor Dam, Bakun Dam, and Kenyir Dam [19].

2.3 Why wetland are needed in a garden?

There are three reasons to add wetland into a garden:

a) An attractive feature for recreation garden:
A garden with wetland can create ecosystems which invite wildlife species such as birds and frogs. Green surrounding with birds and nature sounds can create a calm atmosphere that suitable for recreation activities. People prefer to hang out with family and friends after work in the afternoon to relax and it ultimately can improve their lifestyle. Parents also can educate their children to appreciate nature with exposure to natural surrounding at their young age.

b) Could reduce water bills
By adding a wetland into the garden, and provide a good design to collect water runoff and rainwater, it could reduce the household water bill. Grey water from wetland can feed the garden and slash the water bills.

c) Create wildlife habitat
Other than fish that can live in the wetland, frogs also attracted to the water areas. Frogs are one of the most threatened species on earth, according to the EWT Threatened Amphibian Programme. Mini wetlands could help provide safe havens for endangered frogs. Birds also attracted to wetland as it provides water and also provides food such as fish and aquatic insects in the wetland.

3 Design of Malay Traditional Mini-wetland

3.1 Site selection

As a method to appreciate the wetland, the practice of wetland creation in the garden or backyard is one of the effort to conserve wetland ecosystem. It offers the same benefits as natural wetlands. In Malaysia, the practice of creating mini wetland and appreciating the wetland is still lacked among communities. Wetland contributes a major proportion of fixed carbon in biosphere which ultimately will affect climate if most of the wetland degraded.

A sustainable mini-wetland can be developed in a garden as recreation, education and spiritual site or in a backyard for stormwater management. In order to build a mini-wetland, there are few factors need to consider in selecting a site for a sustainable mini-wetland. Water source and energy supply are basic needed in creating wetland ecosystems. The best spot for wetland development is at the place where the water is readily available, or in existing wet area and drainage area. In a garden, the wetland area should be placed in the lowest plain so the runoff water from rainwater will be collected there. Likewise, rainwater from a downspout or from a paved area can be directed towards the bog garden by a 'dry riverbed' consisting of a plastic-lined channel filled with pebbles. This can be one of the effort to conserve water resource to ensure a sustainable future.

Sunlight-available as energy supply in the mini-wetland area needs to consider to ensure the mini-wetland can undergo chemical interaction such as nutrient supply. Large trees near the wetland can increase the maintenance as the leaves will need to be removed from the wetland to prevent them from fouling the water as they decompose. Safety concern for the wetland also should be considered as it can harm neighborhood such as their properties by excessive moisture and also their children.

3.2 Creating a mini-wetland

Once the location is selected for a site, the shape of mini-wetland must be decided in considering organism that will live in or near the area. Mark out the required shape with string or sand and dig a trench about 1 m deep with sloping sides. To make part of mini-wetland like natural wetlands, the trench of wetland can be line using the clay-lining method. This method used bentonite as the liner as it is partially natural to create a sustainable wetland.

There are few factors need to consider in choosing the plant's type. The homeowner needs to consider the type of vegetation that fits into the rest of the mini-wetland. Plants in or near wetland provide oxygen to the ecosystem, prevent erosion and it helps in increasing the water clarity in the mini-wetland. In some instances, maintaining plants that already adapt in the wetland area is the best way to create a wetland. Usually, homeowners or gardeners used plants that are not adapted to the site or modify the site by introduced new plant species, when it would be more effective to use plants suited to the conditions. The amount of time and interest of the homeowner want to spend and care for the plants should determine what type to install. If the homeowner is not interested in gardening, then select only from the list of basic or low care plants. The more time and care the owner can commit to maintenance, the more variety and color can be designed into the wetland. Numerous landscape plants are well adapted to wet conditions and will provide beauty as well as wildlife habitat.

Select plants that are hardy for the wetland and provide the desired wildlife habitat and aesthetics such as plant marsh or bog-loving plants (bamboo reed (*Elegia capensis*), dekried (*Chondropetalum tectorum*), peacock flower (*Dietes bicolor*), arum lily and the wild iris (*Dietes grandiflora*). Plants that grow in the wetland also need to be identified as it can provide food sources that attract animals, like scabiosa and *Hypoestes aristata*, which attract butterflies. To attract birds, set aside an exclusion area that is far away from the living areas and blocked off from any pets. Plants like water lilies need to be continually flooded. Provide artificial shelters or food sources, such as a feeding platform where kitchen scraps can be placed. Aside from these plants, the wetland will attract a variety of wildlife. Logs, leaf litter, rocks, compost heaps and stone walls can provide shelter for a range of small animals, such as frogs and lizards. The presence of insects, such as dragonflies, and frogs to the wetland is a good indication that the wetland water is healthy.

4 Conclusion

As a conclusion, the value of Malay traditional ecosystem provided by wetland is it is one of the ways of urban communities to improve lifestyle and quality of life, and it increases awareness on protecting local wildlife. The design of Malay traditional wetland can create a sustainable environment in the scope of Malay style. Besides, plants and habitat in the design are introduced, maintained and adapted in our local areas. Lastly, it can give awareness on protecting threatened local species. The idea allows urban communities to practice traditional gardening with the use of local plants in order to preserve heritage species.

Acknowledgement

The authors express their gratitude to the Malaysian Ministry of Higher Education for awarding the Long-Term Research Grant (LRGS, 203/PTS/6727002), the Knowledge Transfer Programme (KTP, 203/PGURU/6750077) and Universiti Sains Malaysia and also all involved in this study.

References

1 J.A. Cherry, Nat. Edu. Know. **10**, 16 (2011).
2 J.B. Zedler, S. Kercher, Annu. Rev. Environ. Resour. **30** (2005).
3 L.M. Cowardin, Carter, Golet, LaRoe, Washington, DC: US Fish and Wildlife Service (1979).
4 M. Finlayson, M. Moser, Oxford, UK: Int. Waterfowl and Wetlands Res. Bur. And Facts on File (1991).
5 W.J. Mitsch, Ecol. Eng., **24**, (2005).
6 W.H. Corner, J.W. Day, Wetl; Ecol. Man, (1982).
7 D.E. McAllister, A.L. Hamilton, B. Harvey, Seawind, **3** (1997).
8 WCMC. Chapman and Hall, London (UK), **585** (1982).
9 Ibrahim, N. Abdul Aziz, N. Abu Hanifah, Pro. Soc. And Behav. Sc., **50** (2012).
10 Clean Malaysia. Available at http://cleanmalaysia.com/2015/11/24/wetland-conservation-in-malaysia/. [Retrieved on 28th November 2017].
11 Millennium Ecosystem Assessment, *World Resources Institute*, (Washington, DC, 2005).
12 M.F. Chislock, E. Doster, R.A. Zitomer, A.E. Wilson, Nat. Edu. Know, **4** (2013).
13 J.M. Olley, *Proceedings of an international symposium* (Alice Springs, Australia, 2002).
14 W.F. Cross, C.V. Baxter, E.J. Rossi Marshall, et al., Ecol. Mono. **3** (2013).
15 G.D. Sherwood, G.A. Rose, Est. Coast. And Shelf Sc. **63** (2005).
16 M.J. Vander Zanden, B.J. Shuter, N. Lester, J.B. Rasmussen, The Amer. Natur. **154** (1999).
17 W.M.A. Wan Mohd Khalik, M.P. Abdullah, The Malay. J. of Ana. Sc., **16** (2012).
18 B.M. Modu, K. Zaleha, F.M. Shaharom-Harrison, Nigerian J. of Fis. and Aqua., **2** (2014).
19 Food and Agriculture of United Nations. Available at http://www.fao.org/nr/water/aquastat/countries_regions/MYS/. [Retrieved on 25th April 2017].

Permaculture Design: Linking Local Knowledge in Land Use Planning for House Compound

Anizah Mohd Salleh[1], *Farah Mastura* Rosli[2], *Norizan* Esa[3], and *Mahamad Hakimi* Ibrahim[2*]

[1]Faculty of Technical and Vocational, Universiti Pendidikan Sultan Idris, Perak, Malaysia
[2]School of Industrial Technology, Universiti Sains Malaysia, 11800 Pulau Pinang, Malaysia
[3]School of Educational Studies, Universiti Sains Malaysia, 11800 Pulau Pinang, Malaysia

Abstract. Permaculture accounts as among the sustainable practices that incorporates diverse aspects of ecology, landscape, organic farming, forestry, agriculture, architecture, and city planning. However, the increasing loss of local values may indirectly influence the sustainability of the system, since local knowledge is an important source of information that influences permaculture design. This study attempts to understand the knowledge of local people in planning and designing a permaculture system subjected to their livelihood, culture and ecological aspects. In particular, this study identifies land use planning for permaculture zones, and design elements in permaculture zones. The study reviews the planning and design element of permaculture that is significant to understand the ecosystem based design with sustainable practices. The results discussed that local people, regardless of their locations either urban or rural, tend to take into account the natural layout, interior architecture and internal activities in planning and designing the zones for their house compound. The zoning is not only beneficial for efficient energy use, but also sustainable both in environmental, social and economic aspects appropriate to the livelihood and culture of local people. Therefore, local knowledge should be considered when designing a permaculture system, as it is likely to design with nature rather than design against nature.

1 Introduction

Permaculture has been recognized as a new ecological paradigm of sustainable practices, together with sustainable development in diverse aspects of ecology, landscape, organic farming, forestry, agriculture, architecture, and city planning [1]. These sustainable practices were adopted both in urban and rural areas as a contribution to environmental protection and acknowledged as ecological design for human settlements. To ecologically design, preserve and maintain natural systems, 12 permaculture design principles considering natural patterns need to be followed [1, 2]. Those are creative usage and response to change; observe and interact; catch and store energy; obtain a yield; self-regulate and accept feedback; renewable use and value; produce no waste; design from pattern to detail; integrate; use small and slow solutions; use and value diversity; and use edges and value the marginal. These 12 design principles can be regarded as a guideline for ecosystem based design, which indicate the ecological processes of planning and designing.

Despite all its merits, conventional sustainable practices adoption with respect to local knowledge is rarely considered in planning and designing processes. The situation has contributed to the failure of sustainable practices that caused by ignorance of nature's pattern [3]. It has led to several disadvantages towards preserving the values of natural systems in sustainable practices, for example, the loss of natural topography beneficial for climatic protection and abundance of plant species rarely consumed by the local people. Other than that, it would be difficult to protect local ecosystem with respect to their local knowledge, and these causes the decreasing of appreciation of local culture among local people. Therefore, the aim of this study is to understand the knowledge of local people in planning and designing a permaculture system subjected to their livelihood, culture and ecological aspects. Two objectives are embedded in this study that is to identify i) land use planning for permaculture zones, and ii) design elements in permaculture zones. These aim and objectives are studied with respect to one of the permaculture design principles which is design from pattern to detail, interestingly focusing on the influence of local knowledge in planning and designing for house compound.

2 Literature review

2.1 The novelty of permanent agriculture concept

The ideas of permaculture, derived from permanent agriculture developed by Bill Mollison and assisted by David Holmgren in 1978, are generating an increasing interest as an approach to natural resources management. This system focused on the ecosystem based design, concerning fundamental factors in ecosystem services of supporting, provisioning, regulating, and cultural [4, 5]. This kind of approach is beneficial to life in all its forms by designing sustainable human settlement as well as preserving and maintaining natural system. In many contexts, permaculture is considered as a holistic approach, especially when it involved permaculture ethics and design principles in planning and designing the permaculture sites as it deals with ecosystem in order to obtain sustainability [3, 6]. Encouraging more environmentally development of ecosystem based design with concern of natural cycle, multi-product, and setting limit for population and consumption, is therefore defined the novelty of permanent agriculture concept.

2.2 Pattern for land use planning and designing

The recurring question of planning and designing permaculture sites is whether and to what extent patterns of land use planning has affected the natural system of permaculture. Natural system clearly involved the ecological processes of human, plants, animal, nutrient cycle, climatic cycle, and weather cycle that harmoniously integrated and complement each other [3]. In line with this, theoretically, planning involves a hands-on or on-site process, while designing is a kind of environmental analysis in which patterns are developed based on the existing situations or structures [1]. Empirically, pattern, in this context refers to the natural features of permaculture sites include topography, soils, hydrography, climatic factor and orientation, vegetation, and wildlife. Therefore, the process of planning and designing a permaculture sites through hands-on or on-site process which consider natural patterns are seen as viable art, science and design development. It is subjected to its advantages in providing greatest impacts and values for the surrounding ecosystem since the planning and designing processes follows the design principles of ecosystem based design. The rationale for patterns in planning and designing permaculture sites is explained through the involvement of landscape by dividing the pattern into zones. Basically, zone planning is a system where elements are designed to be located in certain zoning areas and depends on the regularity of the elements to be used and maintained. For example, herbs to be used in the household kitchen are located in one zone. Another zone may contain the chicken coop, a pond, shrubs and short trees that provide food for the household forming a food forest. Larger and taller fruit trees is located in a third zone further away from the main house while the last zone that may be the furthest away is usually the natural forested area that serves as a supply of firewood and for foraging. Thus different zones require different energy efficiency. In this manner, zones are important for efficient energy planning. Such importance includes designing elements such as trees and plants, animals, and structure and building to be integrated in the zones.

2.3 Approaching permaculture adoption towards local knowledge

Permaculture is an extremely complex system with ecosystem based design for land use and lifestyle. The most obvious relation of permaculture to local ecosystem and knowledge is based on man and nature relationship [7]. Such relationship can be seen through human livelihood interdependence on health and productivity of ecosystem. Natural system may be used positively or negatively depends on how the local knowledge are being used in planning and designing an ecosystem, but human stewardship surprisingly play a vital role in shaping the environment as human is an integral part of nature [8]. In other cases, zones of permaculture should be designed based on natural pattern and necessity of elements step by step through observation, visioning, planning, development, and implementation to avoid harmfulness both to human and environment [6].

3 Methodology

This study is based on the qualitative research methodology by adopting document analysis to identify the influence of local knowledge in planning and designing permaculture system. In details, it is to study how local ecosystem and knowledge could positively contribute to the sustainability of permaculture zones with respect to the livelihood, culture and ecological aspect. The document analysis involves particular review on the current permaculture issues with respect to the local needs, lifestyles, culture, and ecosystem. Importantly, the selection of document analysis is based on the permaculture practices in tropical countries.

4 Results

A total of five (5) zones for land use planning were identified and three (3) design elements were considered in this study. Normally, the permaculture zones are involving zone 1, 2, 3, 4, and 5. The functionality of these zones is subject to the 3 design elements of trees and plants, animals, and structure and buildings. The characteristic of zones and its relationship with design elements is shown in Table 1. The overall land use planning and design elements indicate that the suitability and accuracy of permaculture zones are influenced by the natural pattern of the surrounding. Those are topography, soils, hydrography, climatic factor and orientation, vegetation, and wildlife.

Table 1. Character of permaculture zones and its relationship with design elements.

	Zone 1	*Zone 2*	*Zone 3*	*Zone 4*	*Zone 5*
Location / Size	Near to the building or ¼ acre	1 acre	4 to 20 acre	Any size	Any size
Land use intensity	High intensity: frequently visited	Semi-intensely cultivated	Occasionally visited	Minimal care	Unmanaged
Accessibility	Very good	Good	Good	Poor	Very poor
Design element	Kitchen garden: small trees, rainwater tank	Food production	Large fruit trees Livestock: forage system	Food forest Livestock: self forage system	Timber trees Inspiration Meditation
Energy efficiency	Low energy usage	Low energy usage	Intermediate energy usage	High energy usage	High energy usage

Based on Table 1, the overall operation of permaculture design will influence the energy efficiency of the system. However, the accessibility of the zones will be determined by the land use intensity which symbolized the regularity of the elements usage and maintenance, not the standard immensity of each zone. Due to human intervention such as interior architecture and internal activities of existing building, zone 1 always requires high attention of human activities in order to maintain the functionality of the zone. In contrast, this zone only involves low energy usage as the distance is near to the building, and thus it has very good accessibility as the area is frequently visited subjected to the existence of intensively used as design elements. Therefore, the nearer the zone of the existing building, the highest land use intensity and accessibility, but involved the lowest energy usage. In particular, the most necessary design elements are placed closest to the existing building, which can be easily maintained and accessed by human.

5 Discussion

5.1 Motivation for permaculture

House compound can be seen to satisfy and serve many functions in fulfilling the basic need of human life. The function is reflected by the interior architecture and internal activities of the house, including the natural layout of house compound. The control over house compound was planned by zones and designed by elements through ecosystem based design approach. It is worth noting that efficient energy use can be endorsed to achieve balance and harmony design in response to natural pattern of house compound. In this way, people can save their energy subjected to the high intensity use of the zones and elements. In relating permaculture zones with local knowledge, it is concerning the usefulness of house compound for each zone, for example an area to reflect joyful and aesthetic, route to connect different zones, and space for kitchen activities. Demonstrating efficient energy use was highly important in this study, but meeting the suitability of the space with the right activities and harmony patterns would increased the values of the permaculture design with respect to local knowledge. This may reflect the fact that ecosystem based design has already following the novelty of permaculture in which to design sustainable human settlement and preserve as well as maintain the natural system. In many cases of landscape design, local ecosystem and knowledge do play its role in planning and designing of house compound by adopting man and nature relationship.

5.2 Implications for linking local knowledge in permaculture

Although sustainable practices were important, planning and designing with nature through ecosystem based design with respect to local knowledge did not seem vital for the local culture. Local knowledge was not clearly seen as part of the influential factor in planning and designing for ecological processes. Notably, from this study, there was significant correlation between permaculture design and local values. There is an opportunity to enhance the permaculture values through natural pattern and layout of house compound and change the perception of ecosystem based design is only beneficial for natural system. With the ethical values and design principle of permaculture, the ecosystem based design will have to consider the local knowledge in order to create permaculture system that can be adopted for both urban and rural areas. This is subjected to the fact that permaculture design need to reflect on the livelihood and culture of urban and rural communities rather than only focusing on the ecological or natural layout of house compound as shown in

Figure 1.

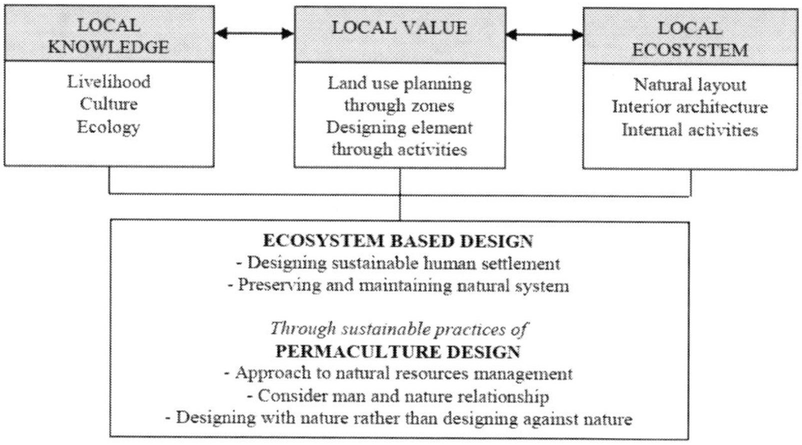

Fig. 1. Relationship between permaculture design and local knowledge, value and ecosystem in planning and designing house compound for ecosystem-based design project.

6 Conclusion

Designing with nature rather than designing against nature represents an ecosystem-based design that illustrates sustainable practices in diverse aspects. This permaculture system is immersed with planning and designing by following the natural patterns of permaculture sites. As such, permaculturist must understand the elements of permaculture in detailed as well as able to find the appropriate pattern of permaculture design. Like other sustainable practices, permaculture is established with regards to the local ecosystem and perhaps it is even more local knowledge scrutiny especially in land use planning and designing. In response, the understanding on how permaculture operates is related to the ethic and principles of permaculture. It is driven from pattern analysis, which constitute visual embodiment of nature's knowledge through observation of our sense and practical knowledge through the pattern as representation of human activity.

Acknowledgement

The authors express their gratitude to the Malaysian Ministry of Higher Education for awarding the Long-Term Research Grant (LRGS, 203/PTS/6727002) and the Knowledge Transfer Programme (KTP, 203/PGURU/6750077) and Universiti Sains Malaysia as well as all involved in this study.

References

1. K. Rothe, Communication +1 (2014)
2. T. Raberg, The Swedish University of Agricultural Science (2007)
3. A. Faiza, A. L. Suleman, S. K. Safdar, S. Farhana, Journal of Environmental Management, **179**, 31-37 (2016)
4. W. V. Reid, H. A. Mooney, A. Cropper, D. Capistrano, S. R. Carpenter, K. Chopra, P. Dasgupta, T. Dietz, A. K. Duraiappah, R. Hassan, R. Kasperson, R. Leemans, R. M. May, T. A. J. McMichael, P. Pingali, C. Samper, R. Scholes, R. T. Watson, A. H. Zakri, Z. Shidong, N. J. Ash, E. Bennett, P. Kumar, M. J. Lee, C. Raudsepp-Hearne, H. Simons, J. Thonell, M. B. Zurek, Millennium Ecosystem Assessment (2005)
5. L. Calvet-Mir, E. Gomez-Baggethun, V. Reyes-Garcia, Ecological Economics, **74**, 153-160 (2012)
6. T. Hemenway, A Guide to Home-Scale Permaculture, (2001)
7. M. H. Nur Huzeima, A. Suriati, *Procedia – Social and Behavioral Sciences*, **42**, 182-189 (2012)
8. P. Bourdeau, Journal of Environmental Radioactivity, **72**, 9-15 (2004)

The Green Management Practice by Local Government

Zikri Muhammad[1,2], *Jasmi* Abu Talib[1], *Kamarul* Md. Shah[1], *Jaharudin* Padli[1], *Bayu Taufiq* Possumah[1]

[1]School of Social and Economic Development, Universiti Malaysia Terengganu
[2]Institut Oseanografi dan Sekitaran, Universiti Malaysia Terengganu

in developing countries face an unprecedented challenge to cut annual management budget without compromising the services quality. Local government top management has been pressured by politicians and public to cut costs but keep up high level quality services. Therefore this study has been analysed the use of green management practise among local government agencies to cut the annual budget and protect the environment. The online services are the main element for green management practise by local government for this study. This study interviewed 630 residents to test the online services provided by local government in Manjung, Perak. Out of the total samples, 92 percent respondents use the online services. Meanwhile, 90 percent of respondents are satisfied with the online services. More than 81 percent respondents agreed that online services can be a medium to share their local knowledge to the local government to cut budget and protect the environment. Besides that, the study also found that, the support by consumer and the internet speed have a significant relationship with the online services usage. This study suggested that, local government should increase the online services to reduce budget and protect the environment simultaneously.

1 Introduction

The Green Management Practices (GMP) is one of the sustainable environmental management. There are several issues of GMP such as policies, procedures, implementation and the public and the private sectors resistance. As we know the rapid urbanization has change the lifestyle of urban citizen, which is lead on the environment damage. To prevent the situation, not only individual, but all the parties including private sector and government agencies must take part to practise the sustainable development. Nowadays, most of developed and developing countries produce the strong policies and procedures on the environment. The GMP by government agencies will create the good image to organization and produce the improve service quality. The GMP by local government will ease the front line service to their customer especially local people. One of the GMP by local government is the use of online services to cut the cost and shorter time. The online services is one of the public infrastructures provided by local government to cut budget and time for high quality services. The public infrastructures provided by local government are the most importance factors that influence the residential satisfaction [1]. Besides that, the GMP through online service will improve the relationship between local government and their customers. Research by Smith show that 82 percent of internet users which represent 61 of all-American adults looked for information or completed a transaction on a government website [2]. Meanwhile, the study by Benton and Simon found that many councils in the USA have already transformed the way they give information and how they manage transactions through online [3]. Nowadays, most local authorities no longer rely solely on 'in-house' operations to deliver either public services or their own internal functions, they need to share their information with other parties' especially with local residents [4]. This can be done through online system where all parties can give information. So, this study will find the level of satisfaction on online services provided by local authority and the readiness of local people to use the online system for dissemination of local knowledge.

2 Methodology

The unit of analysis in this study are the residents of Manjung and Kerian in Perak. This study interviewed 630 residents to test the online services provided by local government. Out of the total samples, 92 percent respondents use the online services. This study measured the satisfaction of online service by local government among the residents' area for urban and suburbs. The measurement covers the online services provided by local authority. The indicators in this study include:

1. E- Rental
2. E-Licensing
3. E-Complaints
4. E- Assessment
5. E- Payment
6. E- Compound (Car)
7. E- Compound (Other)

This measurement used seven-point Likert-type scale ranging from 1 = Very dissatisfied to 7 = Very satisfied for all the quality of online services performance items. This study also will explore the level of readiness of respondent to dissemination of local knowledge by using the online system provided by local authority. For this measurement, this study used five point likert-type scale ranging from 1. Strongly disagree to 5. Strongly agree. To develop comprehensive indicators of online services performance, this research measure total online services provided, level of online safety, method of payment, time consumption, the accuracy of online services and the mobile technology used. A set questionnaire has been developed to get response from customers about the level of satisfaction. The broad aims of the research was to explore the online services performance based on resident's perspective practice by local authority. This study basically used convenience sampling, but taking into account housing site and the different types of housing. This will make sure that the various subgroups in the population are represented. Out of 640 answered questionnaires replied and only 630 were usable. This is representing 98.4 percent of response rate. This study applied the descriptive analysis, factors analysis and correlation analysis to answer the research question and objective.

3 Findings

The majority of the respondents are male with 80.9 percent males and 19.1 percent females. With regard to age, most of the respondent's ages are 41 to 50 years old. In term of marital status, 90.6 percent of the married residents, 1.3 percent is single, and 8.1 percent is divorced. About 78.5 percent had education below or equal to college diploma level compared to other group of educations. The large majority respondents had an income of less than RM5000 per month (95.3 percent) compared to respondents with income more than RM5000 per month (4.7 percent). All the respondents are randomly selected to make sure the sampling bias did not happen.

Table 1: The mean level of satisfaction and Cronbach's alpha value.

Indicators	Item	Mean Level of Satisfaction	Cronbach's alpha
E-Rental	Total online services provided	5.8	0.82
	Level of online safety		
	Method of payment		
	Time consumption		
	The accuracy of online services		
	Mobile technology		
E-Licensing	Total online services provided	6.7	0.85

	Level of online safety		
	Method of payment		
	Time consumption		
	The accuracy of online services		
	Mobile technology		
E-Complaints	Total online services provided	6.0	0.78
	Level of online safety		
	Method of payment		
	Time consumption		
	The accuracy of online services		
	Mobile technology		
E-Assessment	Total online services provided	8.5	0.84
	Level of online safety		
	Method of payment		
	Time consumption		
	The accuracy of online services		
	Mobile technology		
E-Payment	Total online services provided	7.5	0.76
	Level of online safety		
	Method of payment		
	Time consumption		
	The accuracy of online services		
	Mobile technology		
E-Compound (Car)	Total online services provided	6.0	0.84
	Level of online safety		
	Method of payment		
	Time consumption		
	The accuracy of online services		
	Mobile technology		
E-Compound (Other)	Total online services provided	6.5	0.85
	Level of online safety		
	Method of payment		
	Time consumption		
	The accuracy of online services		
	Mobile technology		

4. Reliability Test

According to Nunnally's alpha values equal to or greater than 0.70 can be considered a sufficient condition [5] . This study used Cronbach alpha value for testing the inter-item consistency. The result shows that all the online service by local authority have the Cronbach value more than 0.70 (0.76-0.85). This study concluded that these measures had sufficient reliability.

5 Level of Satisfaction

Generally, this study found that 90 percent of respondents satisfied with the online services. Table 1 shows that mean for level of satisfaction in all indicators for online service more than 5.6. It shows that most of the respondent are satisfied for all online services. This can be concluded that local people are ready to use the technology for their convenient especially payment or complaint services. The use of technology especially online services is one of the GMP initiative by local authority to protect the environment. All the transaction can be done through online system. It will be reduced transportation cost, utility cost, and time-consuming. Besides that, this study also found that more than 81 percent respondents agreed that online services can be a medium to share their local knowledge to the local government to cut budget and protect the environment. They agreed that most of the local services especially for payment and communications' services can be done through online system. Out of 630 respondent 93 percent suggests that local authority should increase the online services for dissemination of local knowledge. Besides that, they can share the local knowledge to the local authority for local planning or land used planning. However, not all the services provided can be use as a channel for local people to sharing their local knowledge.

6 Conclusion

As a conclusion, GMP initiative by local authority by using the online services will protect the environment for sustainable future development. The use of technology can give many advantages not only GMP, it also can increase image and convenient for local people in term of transactions. Most of the local people will enjoy the services from their home or anywhere and at any time to transact with local authority. This GMP will create a good situation for all parties including local authority, local people, private sectors, or any other parties. In this scenario all parties indirectly support the local authority to carry out the GMP for sustainable development. This study also found that the use of technology also can be used for local knowledge dissemination for urban and land use planning.

Acknowledgement

This paper is part of the outcome for the Long-Term Research Grant Scheme (LRGS) program on "Conceptualizing Local Knowledge: A Study on Sustainable Values of Local Wisdom and Indigenous Sciences" Specifically project (203/PTS/6727003) on "innovative Local Knowledge Repository Support System".

References

1. Zikri M, Katiman R. Yusliza M.Y. (2010) Residential Satisfaction with Housing in the Malaysia Context. The International Journal of Interdisciplinary Social Sciences. **5** (2) 379-396.
2. Smith A. (2010) Government Online: The internet gives citizens new paths to government services and information. Sources : http://www.pewinternet.org/files/old-edia/Files/Reports/2010/PIP_Government_Online_2010 _with_topline.pdf downloaded on 19 July 2017A.
3. Benton M. and Simon J. (2016) A Digital Vision of Digital of Local Government in 2025. Public Service Transportation Network. March 2016.NESTA 2016.
4. Sandford M. (2016) Local government: alternative models of service delivery. Briefing Paper Number 05950, 20 May 2016 Source:https://study.sagepub.com/sites/default/files/H%20of%20C%20library%20on%20ousourcing.pdf downloaded on 19 July 2017.
5. Nunnally J.C. (1978). Psychometric Theory. McGraw Hill: New York (1978).

Local Knowledge, Public Policy and Poverty Reduction: A Review on Indonesia Experiences

Bayu Taufiq Possumah, *Zikri* Muhammad, *Jaharudin* Padli, *Jasmi* Abu Talib, *Kamarul* Md. Shah

School of Social Economic Development, Universiti Malaysia Terengganu (UMT)

Abstract. The community based Local knowledge, basically is an adaptation strategies that arise from within the community itself in fixing socially problem related to the life of society itself. Since local knowledge is the result of interaction between the community and the environment, it is needed to help the community living independently. Hence the local knowledge is the core of efforts to reduce poverty that exists and grows in the community and as the big part of from the process of poverty alleviation programs. The complexity of poverty not only can be overcome by purely economic approach, but it is associated with the dynamics of social, political and cultural that inherent in a community. The multi-dimensional problems requires synergy between government policy, community and local knowledge itself. This paper attempts to describes the frameworks of Indonesian policy to link the three sectors to empower the poor exit from poverty problem

Keywords: Local knowledge; Public Policy; Poverty Alleviation; Indonesia

1 Background of Study

As a country with large population, Indonesia known also as the largest democracy in Asia. Challenges for Indonesian government both central and local levels are quite large as well, how far they are able to practice good governance. The right strategy to achieve this good governance is the effectiveness in communicating with people and a way to realizing an effective communication with people is by using local wisdom in governance practices. Local wisdom or local knowledge can be defined as a richness of local cultures containing life philosophy. In Indonesia, local wisdom applies not only locally on a particular culture or ethnic but also cross-cultural or cross-ethnic thereby forming a national cultural values. For example, almost every local culture in the archipelago implements local wisdom that teaches mutual cooperation, tolerance, work ethic, and so on. In general, ethics and moral values embodied in the local wisdom passed down through the generations through oral literature (for example in the form of sayings and proverbs, folklore), and manuscripts

Indonesia has approximately 252 million populations where rice as the main food almost in everywhere. Every year, a minimum of 33 million tons of rice, 16 million tons of corn, 2.2 million tons of soybeans, 2.8 million tons of sugar and 484 thousand tons of beef should be provided. In fact, numbers of poor people is still in a large proportion approximately as 11.4% of the population, thus the problem of increasing rice production and nutrition become a serious attention for the authority. Besides, uncontrolled conversion of 100 thousand ha agricultural land per year cumulate list of challenges to be faced by the government. Other alarming issues to be concerned such are distribution of uneven food, difficult to access the financial resource and information, the lack of use technology and limited access to the markets. Negative impacts of global climate change must also considered in order to increase food production in Indonesia. These are among the factors that led to high levels of poverty in Indonesia, especially in access-less areas to development.

Corresponding author: bayu@umt.edu.my

According to Ritonga (2003), at least there are two important factors which essentially cause the failure of poverty reduction programs in Indonesia. First, poverty reduction programs tend to focus their efforts on the distribution of social assistance for the poor. This efforts difficult to resolve the existing poverty problem since the nature of aid is not to empower but even lead to dependence. Therefore, this aid program cannot be held continuously or regularly. The second factor is the lack of understanding on the causes of poverty itself. The existing development programs are not based on issues of poverty, which caused by various locally factors.

Generally, poverty in a community is similar to what found in other communities. The similar condition does not mean that same approach should be made to tackle poverty in a community. Each community has its own development character that not necessarily same as other communities. Development of community determines problems which occur in both type and dimensions of the society. This corresponds as well with the knowledge, technology, customs, and resources within the community. Thus, it is important to put more attention to the existence of local wisdom employed in a community economic empowerment and poverty reduction program. Therefore, government function essentially to create and implement policies for the welfare, educate, empower, and protect entire society then it is wise to optimize local wisdom in the implementation of government function.

2 Local knowledge and Economic Empowerment

Indigenous knowledge is knowledge generated from community life experiences and as a result of interactions with their environment over millennia. Indigenous knowledge covers virtually every field of human endeavor and usually varies from community to community; one unifying factor being that it is aimed at addressing challenges and solving perceived problems specific to each community (World Bank, 1994). Indigenous knowledge, contrary to widespread dismissal, is not static, but innovative in that it has been adapted over time in response to environmental changes and influences, and as a result of the need for improvement on strategies (Hoppers, 2002). Thus, indigenous knowledge differs from globally accepted knowledge which is often generated from universities and research institutions, acknowledged by the predominant global media establishments and often traces its roots to western knowledge systems (Oguamanam, 2008).

For some post-development writers, indigenous knowledge represents a possible alternative for progress among the world's rural poor. As Escobar (1995, 98) puts it: "[t]he remaking of development must start by examining local constructions, to the extent that they are the life and history of the people, that is, the conditions for and of change". Such approaches can be supported by careful ethnographic work capable of teasing out the complexities of the interrelationships established between communities and places (Herbert, 2000). This implies a change that comes from within communities themselves, having confidence in and deploying indigenous knowledge, among other things, to bring about economic and social progress. There is then a sense of the rural poor having a voice about progress which affects them, and outsiders listening seriously to what the rural poor have to say, learning from them, and respecting their realities and priorities (Chambers, 1983; 2001).

Economic empowerment implies the act of instituting environmental structures to facilitate individuals in bringing change in their social and economic status. Individuals are said to have become economically empowered when they are provided the tools they need in order to work towards self-reliance. According to the Organization for Economic Cooperation and Development (OECD), "economic empowerment is the capacity of women and men to participate in, contribute to and benefit from growth processes in ways that recognize the value of their contributions, respect their dignity and make it possible to negotiate a fairer distribution of the benefits of growth" (OECD, 2011). Thus, economic empowerment if widespread within a society invariably translates to economic growth as strategies that emphasize entrepreneurship has been increasingly accepted and proven to be a veritable channel for the generation of economic growth and innovation across economies, both developing and developed (Acs, Desai, & Klapper, 2008). And in the words of Audretsch (1997). *'Entrepreneurship makes an important contribution to economic growth by providing a conduit for the spill-over of knowledge that might otherwise have remained un-commercialized'*

3 Poverty Reduction and Public Policy in Indonesia

Public or government policy in general defined as everything to do with the government's decision, either program, regulation or others. A policy will run as expected if supported by a various factors, such as the accuracy of policy in line with the public needs and supported by responsibility and participatory of society, as well as compliance with the law and regulations. But as a constitutional state, it is not easy to run the actual function of the law itself without awareness among citizens. This awareness is not only limited to obedience, but in-depth understanding towards the goal of a policy is the essence of a decision, such as the accuracy of decision-making, proper implementation, and supported by various influence elements.

In term of poverty reduction, the Indonesian government strategy through national development applies the *Four Track Strategy*: pro-growth, pro-job, pro-poor and pro-environment. These four track strategies are intended to promote equitable economic growth through the expansion of employment opportunities, poverty reduction, and maintain sustainable growth with an emphasis on environmental sustainability. Social and economic gaps are expected to be well reduced under this strategy. The national development agenda has been included in the medium term development plan (RPJMN) 2004-2009 and 2010-2014. This is also reinforced in the national development plan in RKP 2013 with the theme of *"Strengthening Domestic Economy for the Improvement and Expansion of Public Welfare"*.

In the meantime the government strategy in reducing poverty is focused through three clusters of poverty reduction programs, namely:

- First cluster: Aid program and integrated social protection based on the family which aims to perform basic rights, reduce the expenditure burden and improve quality of life for poor families. The main programs are rice for the poor (Raskin), National health insurance (Jamkesnas), Family Hope Program (PKH), School Operational Assistance (BOS), and Poor Students Aids (BSM).
- Second cluster: poverty alleviation program based on community empowerment through *PNPM Mandiri* program which aims to develop the potential and strengthen the capacity of the poor to engage in the development, improve the income and standard of living through efforts and cooperation to achieve empowerment and self-reliance where community groups or poor districts as the target.
- Third cluster: poverty alleviation program based on economic empowerment of micro and small enterprises (MSEs), which aim to provide wider access to capital source and strengthening the economy for businesses small and micro-scale through business loan scheme (KUR)

By clustering level it is expected the poor families who initially become beneficiaries at first cluster program will be able to meet their basic needs and then move to receive second cluster program to continue to improve their lives in purpose to utilize the cluster program 3 and eventually expected to be free from the bondage of poverty. To strengthen the third cluster program, the government has launched the fourth cluster since 2011 which consist of low price housing program, cheap public transport, Hygiene water for everyone, cheap and economical electricity, as well as life improvement of fishermen and suburban community. For those reasons, allocation budget to meet the development impediments, Indonesian government has determined that the local knowledge and local wisdom should be emphasized and use as a basis of sustainable socio-economic policies. Here, Indonesian government would like to show the importance of recognition and respect for the role of local wisdom and knowledge in the development of economic sector because it has been widely proven that local wisdom when adopted into policy would results better impact.

Rules and policies that are "one size fits all" should be reduced. It is necessary to create, nationally and locally rules which means that the rules apply in the context of smaller scope with local knowledge and local wisdom. This localized policies based on local knowledge and local wisdom. There are diversity aspects in addition to unity (oneness). What occurred during this time is more emphasis on unity, thus it is necessary to emphasize this aspect of diversity also.

4 The Indonesian Government Policy towards Local Knowledge and Community Economic Empowerment

The purposes of community empowerment are to establish self-reliant individuals and communities. Self-reliance means ability to think, decide and do something that is deemed appropriate to solve problems encountered with the use of own capabilities. The purposes of community empowerment: (i) to improve living standards, (ii) to increases confidence, (iii) to provides freedom for everyone to engage the Community Based Development (CBD). There are several main characters of CBD, as follow: CBD resource community based, CBD community participation based. These CBD sustainable community empowerment based can be seen from two points of view: deficit based approach which focus on the problems that exist and efforts to solve the problem, and strength based approach which focus on the potential and capabilities of either community or individual to improve the quality of life. The roadmap of Indonesian in developing local knowledge and community economic empowerment may be seen as following figure:

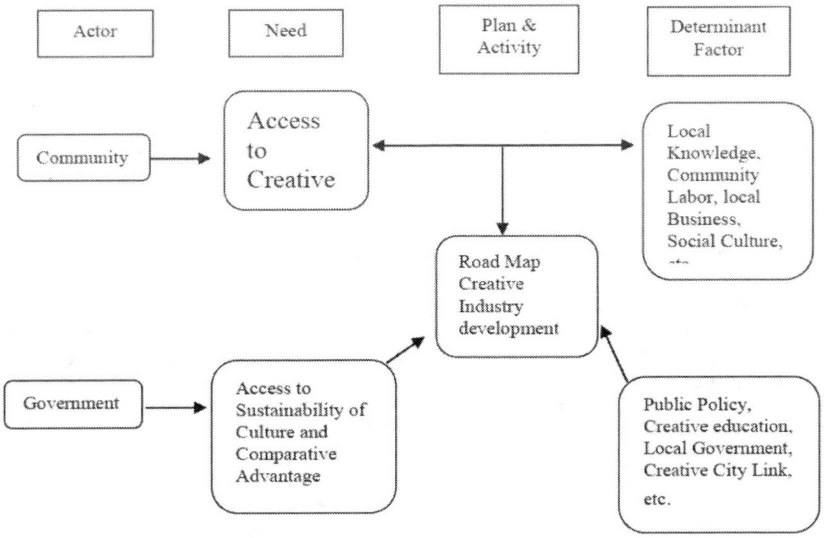

Some of the policies that have been carried out by Indonesian government to make the local knowledge as one of the pillars in the economic and social development, as follow:

4.1 Strengthening The Local Government

Based on the data from *Badan Pusat Statistik*, number of poor people (population with expenditure per capita per month below the poverty line) by September 2016 reached about 27.76 million (10.70 percent) or a decrease of 0.25 million compared to March 2016 which amounted to 28.01 million (10.86 percent). The percentage of poor people in urban areas in March 2016 around 7.79 percent and fell to 7.73 percent in September 2016. Similarly, percentage of poor people in rural areas fell from 14.11 percent in March 2016 became 13.96 percent in September 2016.

Although during the period of march 2016-september 2016 the percentage of poverty as overall has declined, but the number of poor people in urban areas rose as many as 0.15 million people (out of 10.34 million people in March 2016 into 10.49 million in September 2016), while in the rural area fell as many as 0.39 million people (from 17.67 million in March 2016 into 17.28 million in September 2016.

Focusing at the poverty level and the number of which varies in each province, then the effectiveness of poverty alleviation programs cannot be separated from the active role of local government, both at provincial and district/city. Conceptually, the National Poverty Reduction Strategy is important but not sufficient. It is required the active participation of local government and community as well to sharpen the program and the target beneficiaries through the Regional Poverty Reduction Strategy (PRSP) which includes initiatives and local wisdom. Within the framework of the decentralization, central government has delegated broad authority in various fields to the local government, except six major powers (foreign affairs, defense, security and justice, fiscal and monetary, religion). As a unit of government that is closer to the community, the local government is more able to understand and have the ability to implement programs in accordance with the characteristics and needs of communities in the particular region, so that public services can be provided in larger quantities (production eficiency), resources are allocated more effective and accountable (eficiency and accountable allocation) as well as more community participation and the aspirations are more accommodated including the most needy communities.

Those are the reason to give more space to the local government such the rural administration authority to strengthen their role. Eventhough the scale is small and locally which only covers the administrative area of the village itself. Yet the rural administrator has the right and must be participated in determining the direction of national development in general and its own rural development in particular. Rural administration authority formally has to make regulations and able to force the rules legally, but apart from the standpoint of the formalities, each rural traditionally has the wisdom in problem solving. Therefore, the rural administrative authority as a sub-formal institution should not necessarily lead to implement any general operational standards and country's provision. But the rural administrative authority have to adjust wisely and stand along with wisdom and norms that exist in the area differently, therefore every state agency at the bottom level should have flexibility in implementing the policy and governmental functions. This is exactly what lies behind a policy, program or any form of rural authority decisions should not be shaped as rural regulations that bind formally by the positive law of the country. But the rural are has its own traditional norms for any solution to determine the direction of its development, but still expected to play an active role as public servant, facilitator and initiator of solving public problems formally or informally.

4.2 Community Based Natural Resource Management Approach

Community-based nature of resource management (CBNRM) or community based natural resources management approach is a biological strategic management (SDH) in where people can actively participate and play a role in tackling the problems that affect SDH. Thus in this case CBNRM concerns much with the participation of local communities in utilize and maintain SDH in the surround area. CBNRM is an example approach in natural resource management system that considers aspects of justice, equity and welfare of the community around SDH continuously (Jatna Supriatna, 2008). The role of local community in CBNRM is ranging from planning, implementation, monitoring and evaluation. In general, the public role itself determined by three things: 1) the extent to which local knowledge can be appreciated and utilized in forming a good conservation management system; 2) how big local communities concerns about the nature which push towards an effort to preserve and manage natural resources as well as the environment inside and outside the region, and 3) how much benefits (material l and non-material) that can be achieved by society from the conservation area so that its presence has a favorable value continuously. Although there is no success-lesson yet but CBNRM concept might be as reference in building a conservation area management model in favor of public welfare without ignoring the preservation of its biodiversity. Some of issues that are the main target of CBNRM, namely: 1) open access for public (local) and other stakeholders in term of information and management; 2) provide opportunities for people to improve their living standards through the use of available biological resources which will encourage them to continue to maintain its existence; and 3) strengthening the position of the public and other stakeholders in the processes of government policy-making related to natural resource management. To support its success, there are several principles in implementing the CBNRM, namely principle of community development, principle of equality, principle of environmental orientation, principle of respect for the local knowledge/tradition and principle of gender recognition (Jatna Supriatna, 2008)

4.3 Development of Local Food

BPS argues, the role of food commodities to the poverty line is much greater than the role of non-food commodities (such as housing, clothing, education, and health). The role of food to the poverty line in September 2016 pointed at 73.19 percent, that condition was not much different from the situation in March 2016 which amounted 73.50 percent. The type of food commodities with the biggest influence on the value of poverty line in urban and rural areas are rice, tobacco, beef, eggs, sugar, instant noodles, onion and soy beans.

Therefore one of the strategic policies is to achieve the economic independence by enhancing the strategic domestic sectors. One of that strategic sector is agriculture reformation with agenda to improve food sovereignty for community. Food sovereignty is reflected on the regional power to regulate independently that needs to be supported by food security, food policy setting, and able to protect and improve the life of farmers and fishermen as the main actors in that sector.

According to the mandate which set out in *nawacita*, the agriculture ministry has targets to be achieved in 2016 namely the expansion of paddy field to 200.600 hectares, expansion of dry land to 250,000 hectares, rehabilitation of tertiary irrigation to 650,000 hectares, 1,000 units shop for Indonesia farmer, restoration and optimization of 275,000 hectares soil fertility, escort, mentoring and monitoring, constructionof 39 warehouses, better regulation of food imports, improvement farmers capability, development of organic rural farming area.

In the meantime of staple food development, development of local food is also expected to support food sources and consumption diversities in food compliance in Indonesia. Traditional food and specific location are directed to develop a more commercial and to support these steps, a local food processing technologies have been and will continue to be developed to produce fast food. Various types of local food that can be developed such as sago which can be transform into a variety of food/groceries like *papeda, sinonggi, sago rendang, sago lemak, sago lempeng, kapurung, sampolet, gobak sago, and rice analog*. Cassava can be turn into such like *tiwul, enbal, kasuami, kabuto, aruk rice, cassava rice, and rice analog*. Corn can be turn into *corn bose, binte biluhute, kabusunosu, maize*, and *rice analog*.

A variety of things should be done to develop the local food are (1) change the mindset of people that meal does not have to always rice, (2) promoting, campaigning, and socialize intensively for local consumption, (3) improve the image of local food through more practical process, simple presenting, eye catching appearance, (4) cooperating with the *private sector* or *public-private partnership* in the development of local food, and (5) launch a campaign to reduce food waste.

4.4 Development of local culture

Preservation model of local culture through the cultural village need to be developed more with destination management approach. This is because the cultural village that contains culture resources has the opportunity to become a tourist destination. Some strategic steps that can be done based on the tourism destination management (Damanik, Janianton and Frans Teguh, 2013: 30), as follows:

- Leadership and coordination. Strengthening cultural village institutions to carry out management functions of cultural village
- Cooperation and partnerships. Strengthening communication and coordination forum between each cultural village and the relevant local government in order to creates continuous synergies.
- Research and planning. Strengthening development of cultural villages as a form of program, conduct assets inventory periodically and socializing the results of research and develop programs that have been carried out by the provincial government regarding management of cultural village
- Community assistance. The involvement of local government in this regard through various facilitation activities to mobilize all assets held in the cultural village environment include support and participation in the community
- Product development. Strengthening the cultural potential both tangible and intangible that has a sale value as a commodity culture
- Marketing and promotion. Strengthening marketing and promotion through cultural activities for instance cultural competition or cultural festivals regularly

4.5 Muffling the impact of globalization on local Knowledge

1. Re-humanization

Restoring human dignity in the globalization era should be adjusted to the adaptability of the population concerned. The development of religious values, ethics, law, and policy are slower when compare with the development of information and technology. Therefore these issues must be addressed more humanist which means it should be viewed as a whole picture, physically and mentally. Thus development should always lead to the realization of human welfare both physically and mentally. If this is ignored than the destruction of human civilization cannot be offset by re-humanization program therefore all parties should take part and contribute positively

2. Ability to Choose

Every technical thing will be done straight away, does not disputed and filtered based on the values of humanity. By means that if it is supported by moral aspects of religious, social, and related aspects that should determine what may be researched and developed then does not executed otherwise in accordance with the local wisdom.

3. Revitalization

Positive effort is needed to prevent bio-cultural sustainable distortion, because development leads to a new culture in the future. Revitalization will be much affected either positively or negatively by inside and outside factors of the country.

5 Conclusion

Facing the challenges of poverty prospects from local knowledge in the future relies on the utilization and empowerment of local knowledge to manage natural resources and its environment. Knowledge about local knowledge in the communities inheritable from generation to generation, as well as innovation and technology also affect the sustainability of local knowledge in the future. Government as the authority must enforce policies related to the use of adopted technology and diffusion of technologies that protect the environment and natural resources through local knowledge. There are many opportunities for the development of archipelago local knowledge discourse. In addition, local knowledge can be approximated from the values that flourish in it such as religious value, ethical value, aesthetic, intellectual or even other values such like economy, technology and more. Then the treasure of local knowledge become a fertile land to be explored, discussed and analysed in order to increase income and reduce poverty in local communities considering the factor of global developments happened so rapidly. In order to create depth linkages between local knowledge and poverty reduction program there are several important components that need to be intervened are as follows: (1) pro-poor anti-poverty policies, (2) social capital and human resources (improvement of knowledge and skills, social and institutional relations, mutual trust), (3) access to land, (4) the provision of employment, and (5) special attention to women's lives as heads of households (widows) and malnourished children due to cultural factors in the countryside.

Acknowledgement

This paper is part of the outcome for the Long Term Research Grant Scheme (LRGS) program on “Conceptualizing Local Knowledge: A Study on Sustainable Values of Local Wisdom and Indigenous Sciences” specifically project (203/PTS/6727003) on “Innovative Local Knowledge Repository Support System”

References

1. Audretsch, D.B., Keilbach, M.C., Lehman, E., 2006. *Entrepreneurship and Economic Grow*th", Oxford University Press, Oxford, p. 227.
2. Audretsch DB and Thurik AR (1997) Sources of Growth: The Entrepreneurial versus the Managed Economy. Tinbergen Institute Discussion Paper TI 97–109/3, Tinbergen Institute
3. Badan Pusat Statistik (BPS), Indonesia
4. Briggs, J. (2005) The use of indigenous knowledge in development: problems and challenges. Progress in Development Studies 5(2):99-114
5. Chambers R (1983) Rural development: putting the last first. (Longman, Harlow)
6. Departemen Sosial RI. (2006). *MemberdayakanKearifan Lokal bagi Komunitas Adat Terpencil*
7. Herbert S (2000) For ethnography. Progress in Human Geography, 24, 550-568
8. Escobar A (1995) Encountering development: the making and unmaking of the Third World. (Princeton University Press, Princeton NJ).
9. Organisation for Economic Co-operation and Development (OECD). 2011. Women’s Economic Empowerment: Issues Paper. Paris: OECD-Development Assistance Committee.
10. OECD. (2014).
11. Oguamanam, C. (2008). Local Knowledge as Trapped Knowledge: Intellectual Property, Culture, Power and Politics. Journal of World Intellectual Property II , 29-57.
12. Ritonga, H. 2003. Kemiskinan dan Kesenjangan Ekonomi Antar Wilayah. Makalah disampaikan pada Konsultasi Regional Produk Domistik Bruto se Propinsi Riau, di Kepri Juli 2003.

13. Women Economic Empowerment . Retrieved March 3, 2014, from Organization for Economic Cooperation and Development: http://www.oecd.org/dac/gender
14. Zulkarnain, A.Ag., & Febriamansyah, R. (2008).*Kearifan Lokal dan Pemanfaatan dan Pesisir.*, Jurnal Agribisnis Kerakyatan, 1

Nature manifestation in philosophy, diagnosis and treatment of disease based on Ayurvedic and Malay traditional medicine: A comparative study

Wan Faizah Wan Yusoff [1]*
[1] Pusat Pengajian Sains Kesihatan, Universiti Sains Malaysia, Kubang Kerian Kelantan

Abstract. Man and nature is an important concept in the traditional community's universal perspective. This concept can be seen in the universal and cosmological view which discusses questions on the origin of human and nature existence. In traditional medicine context such as Ayurvedic and Malay, this nature manifestation not only can be seen in medical philosophy but also diagnosed medically, as well as treatment given to the disease. This research paper discussion is focused on the community's conception towards natural elements either physically or the opposite and the manifestation of understanding towards their medical philosophy. Among the concept that will be discussed are natural elements such as soil, water, fire, air and ether or better known as the space which is not only important in the context of nature composition but also in the human body based on Ayurvedic medicine. Research is done through the observation of literature and interview with Ayurvedic and Malay traditional medicine practitioners. The outcome obtained has shown that there is a difference between the diagnosis and Ayurvedic treatment which focuses on the characteristics and physical signs compared to traditional Malay medicine which also includes personality diagnostic methods as well as physical attributes and signs for the purpose of disease diagnosis.

Keywords: traditional Malay medicine, Ayurvedic medicine, cosmology, diagnosis, treatment

1 Introduction

The traditional medicine system is a cultural heritage which reflects the thinking of the community for years which is in relations to disease etiology, prevention and treatment method. This medical thought is mostly shaped by the universal view of the community which is also influenced by religion in light of the change of the era just like traditional Malay medicine. In the context of the Malay community, the traditional worldview which also originates from animism and dynamism has faced adjustments with the influence of Hindu-Buddhist and Islam. The worldview of the Indian community is also influenced by

*Corresponding author: faienur@usm.my

elements originating from their faith. Four main life objectives related to Ayurvedic traditional philosophy are *dharma* (religious tasks implementation), *artha* (wealth accumulation), *kama* (fulfillment of desires) and *moksa* (liberation from worldly ties). In this context, optimal health is considered as a prerequisite or *sire qua non* for the attainment of the four objectives.

In Malay and Indian traditional medicine, the relationship between patient, healer and natural elements is truly important and inter-related with each other. This humoral concept and elements and its understanding indirectly shows uniqueness between one traditional system with other traditional medicine system. This research paper will focus on nature's concept in the form of microcosmos and macrocosmos related with traditional Malay and Ayurvedic medicine to show different elements that serves as uniqueness to each of the medication system. Discussion will also cover the diagnosis method and disease classification as treatment methods based on both medication systems.

2 The Concept of Microcosmos and Macrocosmos in Traditional Malay and Ayurvedic Medicine

Nature, according to Malay understanding is divided into two which is microcosmos which is the human as a small nature covering the physical, mind, soul, spiritual, and others as well as macrocosmos which covers the environment and supernatural [6]. In the context of the essence of existence, human is macrocosmos whereas nature is microcosmos since human is placed closer to God. The nature dissident existence in the human body does not refer to the presence of the dissidents physically but only as a symbol which describes the natural manifestation in the human body. In this context, human is the symbol of nature existence as seen from the seven important organs such as the brain, bone, muscle, blood, skin, vein and large vein which describes the seven heavens [8].

Four main elements of the human body based on Malay understanding are blood, *balgham* or mucus, yellow bile water and black bile water which are a symbolic of the four natures. These elements are important in producing temperament which is the disposition and temperament which is unique for each individual as observed in the natural phenomena which arise due to the combination of the four elements. Each element must be balanced in order to avoid imbalance which can next contribute to a disease [8]. The microcosmos and macrocosmos concept is also seen in traditional Indian medicine such as Ayurveda.

According to the teaching of Ayurveda, the humans are a microcosmos which is the universe and himself. The whole universe including the human are made up of five elements which is space, water, fire, wind and soil which does not only form the universe (macrocosmos) and human body (microcosmos). Based on the understanding, the effects of cosmic vibration in the universe are believed to be producing ether or space. The movement of ether produces the air which instigates friction and next, heat. The heat particles produced combines with light to create fire. The heat produced will melt ether elements producing water whereas water that solidifies produces soil or earth molecules. The five elements not only form the universal composition but also shape all other events. Water for instance is in the form of ice which has hard element such as soil. Fire element in ice also melts and produces water. With the help of heat, water produces vapour which has the element of air [11].

Since human is the microcosmos of the universe, based on the views of Ayurvedic medicine, the five elements which shapes the universal composition also dominates the individual self. The element of space or ether covers the space available in the oral cavity, nose, stomach, and blood capillary whereas air refers to body metabolism which works in the digestion system and also brain function which refers to intellect. The element of fire is also believed to activate the retina to enable us to see light. Water is on the other hand manifested in the form of digestive juice, mucous, blood plasma and etc which is an important element in the body. The soil element refers to the structure of the human body which is solid including bones, muscles and tendon [11]. Similar understanding could be seen in the context of Malay traditional medicine.

Based on Malay cosmology, all beings, natural phenomenon as well as food or medicine are formed from the basic element which is soil, water, fire and wind. However, there is one dominant element for each being with two different habits. The earth element for example has two different habits which is wet and cold. Its single attribute only describes God whereas the created beings have two attributes which are coupled and function simultaneously [7]. This combined concept is also available in the Ayurvedic philosophy. *Tridosha* or three basic energy principles or functions that exist in inside human which comprises of *vata, pitta*, and *kapha* is the combination of the five elements. From the spiritual viewpoint, *tridosha* refers to the defects, deficiencies, sins and mistakes in response action, response, processing, integration and disintegration of human with the material world. Therefore, the excessive existence of *tridosha* which comprises of *vata, pitta*, and *kapha* could give negative effects such as disease to humans [9, 11].

Health in the context of Ayurveda and traditional Malay medicine also has a direct relationship with natural elements. The influence and changes of nature such as seasons and cold or hot weather has a direct effect on the human health. Malay *Kitab Tibb* such *Tajul Muluk* and *Bintang Dua Belas* also relates disease and health with the celestial movements [7]. Therefore, the suitable time to undergo treatment is when patient and universe are in harmony. The medication prescribed to patient also has symbols representing certain universe elements in order to create a harmonious relationship between human and nature. This view is also shared by the Indian community which places celestial movement and certain time such as lunar moment and midday as two of the important elements related to bad luck which can also affect treatment. The appearance of certain signs from nature resources such as the sound of owls, the appearance of snakes or the blowing of cold and slow wind which is also believed to be related to bad luck which should be avoided in the process of treatment [9].

3 Diagnosis Method and Disease Classification Based on Traditional Malay and Ayurvedic Medicine

Classification and disease treatment are determined based on the source of the disease. There are two methods in classifying a disease or nosology based on *Kitab Tibb*, *Bustan al-Salatin* and *Taiyib al-Ihsan*. The first method is referring to the diseases related to body parts involved such as the head, sensory organs, abdomen, the lower body part or the skin. These diseases are also divided into certain types based on the symptoms showed for example typhoid which are categorized into fire typhoid, heat typhoid and trembling typhoid. The second method is by classifying the disease whether it is cold because of excessive *balgham* or mucus; heat because of excessive blood humour or related to humour such as yellow fever due to excessive water in the body [7].

Apart from the humour determination method in the body which can instigate sickness, Malay medicine also outlines other methods in identifying the stability or humour changes in the body. This method is based on the examination of pulse movement. Compared to pulse examination in Malay medicine which focuses on the thick, thin as well fast or slow movement of the pulse itself [7], the method used to read the movement of the pulse in Ayurveda is by comparing the movement to movement pattern of animals to show *dosha* which dominates the patient body as shown in the table below [9] :

Movement pattern	Dominating *dosha*
Snake or leech	*vata*
Crow, sparrow, frog	*pitta*
Swan, peacock or cockerel	*kapha*
Monkey, swan	*pitta* and *kapha*
Woodpecker or carpenter	*vatta, pitta* and *kapha*

Apart from the physical-related diagnosis, Malay medicine also shows emphasis on personality methods which involves rituals such as trance, prophecy, and the likes of it using methods or source of specific intermediaries to identify the causes of disease. The knowledge in prophecy especially in astrology has been the culture of the Malays [10]. *Kitab Tibb* and MSS 3290 for example predict the disease which one will contract based on their birth month. Individuals born in the Islamic month of *Rabiul Akhir* for example are said to be epileptic and are physically hot whereas individuals born in *Jamadil Awal* have the tendency to have wind-related disease. The use of intermediary sources such as the Malay weapon, kris that popularized the term 'keris berdiri' or the standing kris clearly shows the beliefs of the community on mystical sources which also help identify future diseases [5]. Aside from kris, other intermediary sources commonly used are kernels which are counted in pairs based on nature element such as soil, water, fire and wind which form the human body. Based on the number of kernels in each part, the shaman will interpret the meaning of elements and next identify the disease [2]. This method shows clear differences in traditional Malay and Ayurvedic medicine.

4 Treatment Method Based On Traditional Malay and Ayurveda Medicine

The concept of Malay medicine regarding medication covers the pharmaceutical material used in treatment and prevention of disease and as a remedy for poison or toxic-based diseases. The intake of certain food or medicine in the context of Malay medicine is for the purpose of changing or influencing humor in the body and not directed to specific organs just like in modern medicine [7]. The medical material used in modern Malay medicine covers quite a number of different types of herbs and roots from nature as specifically stated for the parts needed to be used. These plants were given specific names and most of it could be standardized with the names used today such as white pepper, turmeric, garlic, *jemuju*, white cumin, black cumin, cumin, turmeric, galangal, cloves and others [4]. In light of the humoural doctrin, the medical material in Malay medicine based on *Bustan al-Salatin* and an observation on local practices can be classified into its criteria and health benefits. The medical material such as sour, bitter, stew, slimy, chewy, and watery are categorized as cold and used to treat diseases that are hot in nature [8]. The treatment method is humoral based on alopathy principal or *contrana contranis* by giving food or medication opposite of the disease complications. Diseases categorized as cold such as epilepsy will be treated using heat-based medication such as turmeric, sesame oil and the likes of it to prevent the disease from cold infiltration [7].

Based on the observation of the *Kitab Tibb Pontianak*, the floral choice was made with full accuracy and not only limited to the use of leaves to expose each plant's benefits which also includes fruits, seeds, skin and roots in which each of it has its own therapeutic values [1]. In order to ensure the effectiveness of treatment, medicinal plants are harvested at specific time by giving focus on seasonal changes since plants' active ingredients do not only change according to the condition of the soil and climate but also change according to season. In fact, age and maturity of the plants are also taken into consideration in order to ensure the quality and value of the active ingredients [3]. Apart from botanical, other materials also used in traditional medicine are animals, such as their internal organs, skin, horns and others including minerals such as salt, bauxite, sulphur and other valuable minerals such as gold and silver [1, 4].

Based on Ayurvedic medicinal practice, the main aspects which should be focused on in determining medical materials suitable for diseases are by identifying the causes of diseases, early signs and symptoms based on the law of causes and effects. The selection of medical material in principle is based on the opposite attributes of the disease causes or the disease itself. As for example, the use of *kuinin* is believed to be the best medication for malaria. Similar method was accepted for use in homeopathy medicine. The single or combined use of medicine also shows similarity with traditional Malay medicine as discussed earlier on in modern medicine [9].

The source of Ayurvedic medical material comprises of plants, animals and minerals such as that in Malay medicine. According to Ayurveda, medicinal plants can be divided into two types which is cold or hot in nature based on the taste which was also discussed in Malay medicine. Medicinal plants which are categorized as hot in principle has a strong, sour and salty taste whereas cold medicinal plants has a sweet and bitter taste. Each taste, according to Ayurveda has its own benefits in treatment of disease. The sweet (*madhura rasa*) for example is good for the health of the brain, heart, throat and skin apart from being able to treat burns and quench thirst. The sweet taste can also increase milk production, body growth and skin condition. However, if taken excessively, it will give a negative effect such as the formation of fat, increase of weight and causes laziness as well as the loss of appetite, weaken the digestion and causing diseases such as coughing, fever as well as cold [9].

Apart from the changes in food intake and balancing excessive *dosha* in order to avoid a disease, Ayurveda also emphasized on food adjustments according to seasonal suitability to avoid seasonal change disease or the 'Three Season Diet'. Based on the observation made in winter for example, animals such as squirrels take a lot of protein such as beans. Therefore, the intake of the same protein is also encouraged in Ayurveda in order to maintain health in winter. During fall, one is encouraged to take green plants and seasonal fruits such as berries to avoid allergic-based disease and the accumulation of mucus which happens a lot during that season [9].

The influence of seasons and specific time are also emphasized in Malay medicine. According to the MS33 *Kitab Tibb*, the wind humor is dominant in summer whereas blood humor is active in moderate season. Wet or monsoon season shows that *balgham* or mucus is more dominant. Apart from that, there are also certain times in a day when some humour loses its balance and may cause diseases. Morning time especially shows a tendency to bring the wind humour disease whereas the evening is related to the *balgham* or mucus humour-based disease. In the night time, there is also a tendency for diseases caused by the imbalanced of wind and *balgham* or mucus. The day shows diseases caused by the imbalanced of blood and wind humour. Therefore, in facing certain seasons, the intake of suitable diet is encouraged in order to ensure optimal health [7].

5 CONCLUSION

Malay medicine shows knowledge and local wisdom based on the accumulation of knowledge and experience as well as interaction with different culture. Research on the essence of Malay medicine manuscript or *Kitab Tibb* is important because it includes different viewpoints compared to what has always been composed by the orientalists aside from showing the scientific sides of Malay medicine. Although there are differences between the diagnosis and treatment in traditional Malay and Ayurvedic medicine, in principle, there are similarities in certain health concepts such as the context of human relationship self with the nature which shows a mutualism symbiosis that cannot be separated. The manifestation of nature in Malay and Ayurvedic medicine is clearly shown not only in diagnosis but also in its holistic and unique nature of treatment compared to modern medication which only focuses on certain aspects, separating individuals not only from himself as a whole but also from his relationship with the community and nature. The knowledge of Malay and Ayurvedic medicine which has been used and tested on its effectiveness for decades has seen a focus on the inter-related relationship between human and nature as well as adjustments to the environment in order to create harmony and balance. The relationship between physical, spiritual, environmental, psychology and health in Malay and Ayurvedic medicine shows that knowledge and local wisdom are not only high in value but also a source of amazement.

References

1. Abdul Razak Abdul Karim, *Analisis Bahasa dalam Kitab Tib Pontianak* (Dewan Bahasa dan Pustaka, Kuala Lumpur, 2006)
2. Aishah Mohamed dan Aziz A. Bidin, GMJSS, **8,** 4 (2012).
3. Faisal @ Ahmad Faisal Abdul Hamid dan Nurul Wahidah binti Fauzi, Jurnal al-Tamaddun, **7** (2012)
4. Harun Mat Piah, IMAN, **3,** 3 (2015).
5. Mohd Farhan Md. Ariffin, Khadher Ahmad, Nor Izati Samsudin, Mohamad Zaim Ismail & Muhammad Ikhlas Rosele, GJAT, **3,** 2 (2013).
6. Muhammad 'Uthman El-Muhammady in Yaacob Haron (Ed), *Pemikiran Melayu dalam Kosmologi Melayu*, p. 10-22 (APMUM, Kuala Lumpur, 2001)
7. Pisol Maidin, *Konsep Kesihatan dan Etnofarmakologi: Satu Kajian Falsafah Berdasarkan Beberapa Naskhah Manuskrip Perubatan Tradisional Bustan al-Salatin, Kitab Tibb dan Taiyyib al-Ihsan* (Unpublished dissertation, Universiti Malaya, 1995)
8. Pisol Maidin in Rahimah A.Hamid, Mohd Kipli Abdul Rahman & Nazarudin Zainun (Ed), *Kearifan Tempatan: Pengalaman Nusantara Meneliti Khazanah Sastera, Bahasa dan Ilmu*, p. 180-192 (Penerbit USM, Penang, 2013).
9. Sharma, S.C, *Ayurveda the ultimate medicine* (Wisdom Tree, New Delhi, 2005).
10. Syed Haashim Syed Abdul Rahman *Kosmologi Kitab Tibb Melayu: Satu Penghayatan Awal*, Proceeding of Seminar Sains dan Teknologi Malaynesia (2017).
11. Vasant, L. *Ayurveda the science of self-healing: A practical guide* (Lotus Press, Wisconsin, 1990).

SHS Web of Conferences **45**, 05002 (2018) https://doi.org/10.1051/shsconf/20184505002
ICLK 2017

Postnatal Care Practices among the Malays, Chinese and Indians: A Comparison

Zuraidah Mohd Yusoff[1], *Asmiaty* Amat [2], *Darlina* Naim[1] and *Saad* Othman[1]

[1]Universiti Sains Malaysia, Penang, Malaysia
[2] Universiti Malaysia Sabah, Kota Kinabalu, Sabah, Malaysia

Abstract. In Malaysia, each race has its own traditional medicine practice which has existed for hundreds of years before the coming of modern medicine. Also, each race has many kinds of practices that had been around maintaining the health care of the respective community. All of these races or ethnic groups regard that it is very important for new mothers to be nursed back to health and thus each has its own specific and special postnatal or maternity care. The treatment during the postnatal or confinement period is generally considered to be good and safe and can help the new mother to gain back her health to the pre-pregnancy status. It is also belief that the ingredients used are natural and usually do not caused harm to the mother's condition. Hence, this paper is the result of the study on the traditional postnatal care practiced by the Malay, Chinese and Indian communities in Malaysia. This study was conducted through interviews and review of literature. The results obtained showed that there are a variety of treatments and practices during postnatal or confinement period for each of the race. In addition, traditional postnatal care during confinement are still being sought after and followed by the different races in Malaysia.

1 Introduction

According to the World Intellectual Property Organization (WIPO), traditional knowledge is a living body of knowledge, skills and practices developed by the indigenous or local communities and are sustained as well as inherited from generation to generation within a community. This often become part of its cultural and spiritual identity. One of this knowledge is in the practice of traditional medicine. World Health Organization (WHO) defines traditional medicine as health practices, approaches, knowledge and beliefs incorporating plant, animal and mineral based medicines, spiritual therapies, manual techniques and exercises, applied singularly or in combination, to treat, diagnose and prevent illnesses or maintain well-being. Hence traditional medicine based on traditional knowledge is a medical practice that has long existed in the community prior to the introduction and influence of modern medicine. Currently, traditional medicine is growing worldwide and has become one of the most important branch of medical science giving many impacts towards improving the health of the community. Traditional medicine of course also reflects the background and the origin of a culture. In other words, even though modern medicine has reached the era of globalization, traditional medical practices are not being side-lined or neglected but in fact are still being in demand.

Malaysia is a country with multiracial population with Malay, Chinese and Indian constituting the majority of the population. Each of these races is rich in traditional knowledge and practices that has been in place for centuries. Thus the diversity of races in Malaysia has further enhanced the practice and health care of the country.

One important area of the traditional medicine is involved with the birthing practice observed by women. The traditional birthing practice is defined as all beliefs, behaviours and rituals done during the ante-natal, labour and birth and the post-natal period. These practices are socially constructed and shaped by the perception and practice of the culture. Hence many cultures or ethnicities around the world practice specific postnatal rituals or treatment. This is said to avoid ill health in later years even though at times the immediate effects are not felt. These postnatal practices or rituals would permit the mother to be pampered for a period of time after giving birth. Some of the rituals includes defined period of rest, hygiene practices, food restriction and recommended, special treatment for the mothers as well as breastfeeding and infant care. However with modernization and globalization these post-natal rituals might not be practiced or followed. In Malaysia, the Malays, Chinese and Indian communities do have their own belief and practice in the postnatal care. Thus this paper will focus and describe the postpartum practices in these three major ethnic groups and to identify the similarities and differences in this traditional postnatal belief and practice.

2 Maternity Confinement

Maternity Confinement is one of the traditional treatment which has become a tradition for women who have given birth to a baby. The purpose of confinement is to heal and restore energy and health within a specified period of time. Each race has its own unique methods of post-natal care be it the Malays, Chinese or Indians. Although the Chinese in Sabah and in Peninsular Malaysia are of the same ethnicity however the practice during confinement is a little different because the culture and practice of the Chinese community in Sabah has assimilated with the indigenous communities of Sabah.

3 Midwifery Practice Among Malay, Chinese And Indian

The Malay midwife is called a 'bidan' who traditionally play an important role in the Malay women health. In obtaining more information on traditional Malay medicine and postnatal care specifically, interview was conducted with a Malay traditional medicine practitioner who is still actively practicing traditional Malay midwifery. The 'bidan' is 62 years old, from Teluk Kumbar, Penang and has been in the field of midwifery for about 27 years. According to her, the midwifery skills she acquired are inherited through generations from her mother side. While the information regarding the confinement practice among the Chinese was obtained from the interviews with two Chinese women who have just given birth together with the midwife who took care of them during labour. This Chinese midwife is called 'Yuet Pui Pok' or "Pui Yit" or confinement nanny which literally means companion grandmother. The information relating to postnatal practices among the Indian was obtained from an interview with an Indian lady from Penang.

According to the Malay midwife, care of women after childbirth is very important for maintaining the health of the women so she would recover and be back to the normal status as well as to maintain the health of the reproductive system of women after childbirth. Midwifery or postnatal care can only be started one or two weeks after the childbirth. This ensures the stitches during childbirth or labour have healed to avoid any other complications that may occur.

For the Chinese, they will recruit and hire a trained confinement lady (companion grandmother) to take care of them during the process of confinement The confinement lady is paid not only to take care of the new mother but will also take care of the new born for a certain time period. The confinement lady will take care of the new mother's bathing needs as well as her diet for 30 days. Hence for this, the confinement lady will be paid with a modest salary. Nowadays maternal care institutions have been commercialized and extremely well received in the Chinese community. Although it looks quite modern and with efficient management, postnatal or maternity care practices still follow traditional features.

The traditional Indian postnatal practices is to help the mother to revitalise her body by making it warm again. The belief is that blood lost during labour had caused the body to go into a cold stage. The confinement period is for 30 days whereby many special practices are being followed. For those families who could afford, a special assistant or maid would be employed that looks after the mother and the baby. Otherwise, an elderly member of the family would come to help.

4 The Malay Postnatal Care Or Confinement.

According to the WHO, the postpartum period starts about 1 hour after the birth of the placenta when immediate care of the mother and the infant starts. Although not officially sanctioned, traditionally the postpartum period is supposed to end 6 weeks after birth. The period of 6 weeks fits very well into cultural traditions in many countries, where often the first 40 days after birth are considered a time of convalescence for the mother and her new born infant. The post-natal or post-partum or confinement period observed by the Malay women is usually 44 days although a longer period of 60 days or 100 days are followed by some. Special precautions are followed for those women who had undergone Cesarean (surgery) birth as compared to those with normal birth.

Among the Malay community, the usual postnatal treatment or midwifery care involved during the confinement period nowadays are hot compress (bertungku), heat treatment (berdiang), herbal bath, body wrap (berbengkung/berbarut), and body massage. Also during this confinement period, food that are considered as cold, 'windy' and poisonous are prohibited. Examples of prohibited food are cucumber, long beans, bamboo shoots, pineapples, crabs, and certain types of fishes. Hence food which are categorized as hot are consumed. The women are also encourage to consumed tonic drink (jamu) or herbal paste (makjun) made from herbs which is considered as hot.

4.1 Hot Compress (Bertungku)

Hot compress is a form of point massage using heated objects. There are two types of 'tungku' (apparatus used for hot compress) that could be use i.e. iron 'tungku' (club-like iron with a handle) or stone 'tungku' (sphere-shaped river stone). These 'tungku' are of varying weights and the weight used would depend on the size of the woman's body and usually the two kilogram 'tungku' is being used. If a woman is an obese, a heavier 'tungku' tends to be more suitable for optimum impact. To prepare the hot compress, the 'tungku' is heated up and then placed on some herbal leaves such as *Alpinia galangal* (lengkuas) *and Morinda Citrifolia* (mengkudu) before being wrapped in a cloth. Then using specific technics the hot compress is being applied to areas of the body such as the abdomen and thigh applied. Traditionally, the hot compress is applied for at least half an hour to an hour

4.2 Heat Treatment (Berdiang)

The treatment is performed using a brick wrapped with a variety of herbs such as piper *sarmentosum* (kaduk) leaves, *Erythrina fusca* Lour leaves (dedap), *Piper betle* leaf (sirih), lemongrass and ginger. The bricks are heated before being wrapped with the leaves and a piece of cloth. It is then placed beside the body of a woman i.e. approximately two inches from the body. This process can be done while the woman is lying down or when she is asleep. According to the Malay midwife, the process is very effective in removing toxic substances from the woman's body through her sweat.

4.3 Herbal Bath

For some of the Malays, there are practices that do not allow mother after childbirth to bath for the next three days. However after the childbirth, the mother's body would be wipe with warm water for blood to flow better. However, this is not observed by some whereby they are allow to bathe a day after delivery. Woman in the confinement period is also advised not to bathe in the early morning as to avoid the 'wind' from entering the body and causing body aches and pains. During the herbal bath, the temperature of the water need to be warm to hot (preferably hot) and with various types of herbs such as *Desmodium gangeticum* (daun Meringan) added. At the same time, simple massage in the water is performed on the woman's body so as to soften all the taut and strained muscles during the process of giving birth.

4.4 Whole Body Massage

The type of massage after childbirth is the deep tissue and therapeutic whole body massage and is performed by the Malay midwife who is an experienced masseur. The part of the body to initiate the massage would depend on the midwife. As for the midwife interviewed, the massage technique would starts from the top of the head and ends at the feet. Massage therapy is fundamental to accelerate blood flow because it is believed that after giving birth, the blood flow in the woman's body slows down and will cause her to experience body pain and numbness. Correct and suitable massage technique is very important so as to get the best effect. The most important part of the body is in the nerves and muscle of the thigh, hip and the back. Usually the mother could opt for various number of massage sessions however a minimum session recommended is three.

4.5 Body Wrap (Berbengkung/Berbarut)

Barut or body wrap or abdominal binding is another common treatment performed during postnatal care. This is done by smearing or spreading an herbal paste made from specific herbs and spices on the entire abdomen before binding was done. Tight wrapping or binding is done by using a piece of long cloth (corset like) covering the abdomen area starting from just beneath the breast to the thighs. The size of the barut would depend on the woman's body size. This treatment aims to restore and deflate the woman's abdomen after child-birth and to regain the mother' original body posture and shape.

4.6 Vagina Steaming (Tangas)

Vagina steaming or tangas in Malay is a traditional method that uses steam derived from herbal mixture. This tangas treatment is given with the woman sitting on a wooden chair that has an opening in the middle of the perineum area thus exposing to the steam or vapour coming from a bucket of steaming water with herbs in it. This tangas treatment is intended to lift up the uterus back to its original position and also to shrink the uterus which has expanded during pregnancy. Tangas would also make the body feels lighter and comfortable. This is usually performed in the final phase of the 44 days confinement.

5 The Chinese Postnatal Care Or Confinement

For the Chinese community, the confinement period observed by woman who had given birth depends on the individual but usually it spans a period of one month (30 days) or up to 44 days as the duration observed by the Malay woman. For the Chinese women who really adhere to the practice of confinement, they are not allowed out of the house for the entire confinement period.

The basic principal the Chinese woman most adhere to during this confinement time is keeping the body away from colds (wind). As a result, the woman in confinement would not bathe for several days or weeks (depending on her endurance, namely comfort). It is also believed that bathing would cause evil 'wind' entering the body and causing joint pain. After several weeks, the mother will be bathe with water that has been boiled and let cooled down (relatively warm temperatures). The hot water need to cool down on its own and cold water is not allowed to be mixed with hot water so as to cool it. The bath water is then usually mixed with the leaves or roots of bitter melon or bitter gourd (*Momordica charantia*). Herbs for the bath is also easily available and sold in the shops of traditional Chinese medicine. Chinese woman in confinement is strictly not allowed to wash her head for fear she will be hurt as a result of the cold. In Sabah, the Chinese and Indigenous woman specifically use water mixed with the herb 'sambung-sambung' (***Blumea balsamifera)*** the confinement bath. This herb is said to be good in preventing the body of the woman in confinement from wind (cold) and thus avert from body aches. There is also a practice of the Chinese community not allowing the use of fan for 30 days. However, this matter is seldom practiced by modern society.

In addition to bathing care, the Chinese are also very concerned with food and nutrition during the confinement period. During this period the woman is not encouraged to drink lot of plain water instead the decoction of dried longan and red dates are preferred. The Chinese society also emphasized on taking herbs such as ginseng soup, *red date* and 'lihing' during confinement. According to the midwife, the Chinese community in the state of Sabah practice drinking 'lihing', which is a type of liquor made from fermented rice and cooked with chicken and ginger. This soup is consumed so as to get rid of the 'wind' in the body and to warm-up the body of mother who had given birth. As for the Chinese community in peninsula Malaysia, the food for woman during confinement are specific fishes that are considered not harmful and do not cause itching such as pomfret, red fish and white fish cooked with sesame oil and ginger. While the vegetables are very important in the diet, Chinese society does not encourage the consumption of vegetables during confinement since most of the vegetables are considered cold and cause the body to become weak and caused joint pains. For the Chinese postnatal care, body wrap (berbarut) and hot compress (bertungku) are not usually performed.

In addition, emphasis is also given to the new-born. As an example, in Sabah eucalyptus oil is applied to the whole body of the baby before and after bathing. According to the practice of Chinese community, the baby will be bathed with a mixture of herbs such as the roots and leaves of bitter melon (*Momordica charantia*) because it is believed to prevent the baby from jaundice and trapped wind.

6 The Indian Postnatal Care Or Confinement

The postnatal practices among the Indians are rather similar to those practices among the Malays. After delivery the woman or new mother is allowed to bathe but only with warm water that has been boiled with herbs such as neem (*Azadirachta indica*), mengkudu (*Morinda citrifolia*) and lemuni *Vitex Trifolia* var Negundo) leaves. However, washing the hair daily is not allowed in order to prevent cold. Hair washing is done on odd number days such as day 3, 5, 7 until the 1month confinement is over. The hair need to be dried quickly since it is belief that leaving it wet can cause post-partum depression. The woman is not allowed to bath too early in the morning but can only do so after sunrise or mid-morning. She is also required to keep her body warm at all time. During bathing, body scrub (lulur) made from herbs including turmeric is applied.

Similar to the Malay and the Chinese, the Indian also observed some food restriction during the confinement period. Food that are considered 'cold' or 'windy' should be avoided. Examples of these food are cucumber, murungai (*Moringa oleifera)* fruits, eggplant, jack fruit and grapes. However, some food are encourage such as shark meat and boiled fenugreek seeds are said to increase milk production. Spices such as black pepper, cardamom, aniseed, turmeric, ginger are also encouraged to be added to the food. The woman is given herbal concoction (makjun) to be consumed during the confinement period. The woman is allowed to drink warm water in moderation but ice cold water is prohibited.

Just like the Malays, the Indian postnatal care or practice also apply massage or *maalish* to the new mother. This is done daily by using oil such as neem oil, mustard oil or coconut oil. Among the Indian, postnatal girdle (bengkung) is also used. This is done by applying the oil on the mother's abdomen and binds it tightly with a long cloth so as to tighten up the flaccid muscle. Hot compression is not done in postnatal care among the Indian.

Right after delivery, the new-born will be whispered with the name of the Hindu god. At 16 days old, special prayer is done for the new-born at home and at 30 days the new-born will be taken to the temple for prayer to be recited for him/her. Usually food would be served to family members and friends attending the occasion. The new-born also undergoes daily massage with oil being applied to the entire body. It is thought to be good for the baby's bone and overall development.

7 Conclusion

There are similarities and differences in the practice of postnatal care among the Malay, Chinese and Indians. The similarities are the practices in ensuring recovery and restoring the health of women after childbirth back to non-pregnant status in a defined confinement period. The basic factor in the postnatal care that was greatly emphasized by the three races is the fear of the 'wind' and cold which can caused illnesses such as body aches, fever and bloating in the woman which would hinder recovery. Food restriction is being observed in all the three races. The slight differences observed in the concept and practice between the three races is due to the different culture inherited from the elderly.

References

1. Anisah Barakbah (2007). Ensiklopedia Perbidanan Melayu. Sebuah Perbendaharaan Ilmu
2. Perubatan dan Penjagaan Kesihatan. Utusan Publication & Distribution Sdn. Bhd.
3. Azuzay Zamani (2001). Traditional Practices In Postnatal Care: The Malay Community In Malaysia. Trinity Student Medical Journal . 2. Pp 30-31
4. Cindy-Lee Dennis, Kenneth Fung, Sophie Grigoriadis, Gail Erlick Robinson, Sarah Romans & Lori Ross (2007). Traditional Postpartum Practices And Rituals: A Qualitative Systematic Review. Women's Health . 3(4), pp487-502
5. Eliana Nasir, Sandra Mackey, David Arthur, Piyanee Klainin-Yobas, Helen Chen and Debra K. Creedy (2012). An Exploratory Study Of Traditional Birthing Practices Of Chinese Malay And Indian Women In Singapore. Midwifery 28,e865-e871
6. Indian confinement practices https://www.babycenter.com.my/a1042118/indian-confinement-practice.
7. Lenore Manderso. (1981) Roasting, Smoking And Dieting In Response To Birth: Malay Confinement In Cross-Cultural Perspective . Soc, Sci & Med 15B, pp 509-520
8. M Hishamshah, M bin Ramzan, A Rashid, W Wan Mustaffa, R Haroon, N Badaruddin (2010) . Belief and Practices of Traditional Post Partum Care Among a Rural Community in Penang Malaysia. The Internet Journal of Third World Medicine. 9 (2). pp1-9
9. Poh Bee Koon, Wong Yuen Peng & Norimah A. Karim (2005). Postpartum Dietary Intakes and Food Taboos Among Chinese Women Attending Maternal and Child Health Clinics and Maternity Hospital, Kuala Lumpur Mal J Nutr 11(1), pp1-21
10. Samad. A. (2005). Warisan Perubatan Melayu. 2nd Edition. Dewan Bahasa & Pustaka, Kuala Lumpur.
11. Siti Zubaidah, Norsuhana Ah dan Fatan Hamamah (2012). Penggunaan Haiwan bagi Perubatan Tradisional dalam Kalangan Masyarakat Pribumi di Asia: Satu Ulasan. Pahang: Universiti Sains Malaysia.Unschuld P.U. 1985. Medicine In China: a history of ideas. United States of America : University of California
12. WHO, Geneva (1998). Safe Motherhood . Postpartum Care of the Mother and Newborn: A Practical Guide
13. WHO (2003). Traditional Medicine. WHO, Geneva. SSN 1393-9572. Kuala Lumpur
14. World Intellectual Property Rights (WIPO 2015). Traditional knowledge and intellectual property – background brief. Retrieved from http://www.wipo.int/pressroom/en/briefs/tk_ip.html.
15. Wong KC & Wu L-T. 1932. History of Chinese Medicine. China: The Tientsin Press

NGALAI RITUAL : A COMMUNAL HEALING PERFORMANCE

Mohd Kipli Abdul Rahman[1], *Nurul Farhana* Low Abdullah[2], *Solehah* Ishak[3]

[1]Universiti Pendidikan Sultan Idris (UPSI)
[2]Universiti Sains Malaysia (USM)
[3]Akademik seni budaya dan Warisan Kebangsaan (ASWARA)

Abstract. This study delves into the *ngalai* ritual performance practiced by *Bajau Samah* community in Kota Belud, Sabah, Malaysia which serves as a healing method. The study was done using a qualitative approach through the ethnographic strategy application concentrating on the relationship between behaviour and culture. In terms of performance, *ngalai* is a combination of the elements of dance, singing, music and acting. Traditionally, the *ngalai* ritual is performed for healing purposes that is based on the animistic religious beliefs which is derived from believe in the existence of spirits (metaphysical entity). Individual who is able to see spirits is said to be capable and have the privilege of curing a disease caused by *tulah* (plague) due to the interference of spirits. "The community of *ngalai* practitioners believe that the physical world is influenced by the metaphysical world, especially where it concerned the existence of other worldly beings, the myths and spirits associated with their ancestors. They also believe that the places they inhabit are also shared with the presence of these other worldly, spiritual beings. Based on these beliefs, offerings are prepared and rituals are done in the *ngalai* performance with the main aim of respecting and asking permission from these spiritual beings to help with the healing process." The healing aspect in *ngalai* is categorized as communal healing as well as "faith healing", referring to "the ritual practice that relates to religious beliefs and healing" that happen following the method adopted involving patients and the public. Thus, this study seeks to present a justification of ngalai ritual performance that serves as a medium of healing and is still significant in the present context.

Keywords: *ngalai*, ritual, *Bajau Samah*, *tulah*, communal healing, faith healing

1 Introduction

Kota Belud is a small district located in the northwesterly coast of Sabah (East Malaysia), 70 kilometers off Kota Kinabalu, the capital state of Sabah. The name Kota Belud is believed of have originated from the Bajau language, which is 'kota' meaning the fortress while 'belud' means 'hill'. Thus, Kota Belud is defined as 'a fortress on the hill'. According to narrations, years ago, during war, the people of Bajau made hills as their shelter to rescue them from enemies. For a long time, there were various people who inhabited the district of Kota Belud namely Bajau Samah, Kadazan-Dusun, Cina, Irranun, India and Ubian. However, the people of Bajau Samah constitute the largest population. Up until today, the population of the district of Kota Belud has reached 94,000 persons.

From this, the Bajau[1] people are generally the second largest group in the state of Sabah. Before the reign of British in the 19-th century, the people of Bajau were nominally placed under the ruling of Brunei and Sulu. The

[1] There are various types of Bajau in Sabah, namely Bajau Samah, Bajau Marudu, Bajau Ubian, Bajau Tuaran, Bajau Suluk as well as Bajau Laut. The differences are apparent through the district, their origin and customs. However, the people of Bajau remain in the same society and family.

Bajau people were characteristically maritime, where their skills and knowledge on the sea were never doubted. They were seafarers and lived nomadically, and their source of sustenance depended on sea life. Nevertheless, the Bajau started to move to the land when there were economic activities and they pursued the yields of the forests to survive. For this reason, they selected Kota Belud as their land location until the district was inhabited by the majority of the people of Bajau Samah compared to other districts. The choice of this location as the residential area of Bajau Samah is because the location would be suitable for agricultural-based economic activity.

2 Bajau Cosmology

"Universally, the term cosmology is referred to the study of cosmos. Specifically, it is a study that involves complex issues regarding origin of the universe, i.e cosmogyny (Juli 2001; Singaravelu 2001; Hood 2001) and all aspects that related to world order including the existence of supernatural world and beings as well as sociocultural aspects because human beings are also part of the cosmos. Linked to this term is world-view which is defined as a cognitive-based perception or idea about the existence of something, and also concept about attributes, individual self as well as his or her social surroundings (Geertz 2008: 135). Dilthey also defined world-view or *weltanshauung* as "complex of ideas and sentiments comprising i) beliefs and convictions about the nature of life and the world ii) emotions, habits and tendencies based on these iii) a system of purpose, preferences and principles governing action and giving life unity and meaning" (Symonds 1991:15). The rituals and traditional practices in death customs are often associated with cosmology and world-view because the belief and practices related to death can provide a window for viewing a society's social organization, cultural values and world-view "(Gang Chen 2000).

According to Halina Sendera Mohd. Yakin. (2013), the "Bajau cosmology seems to have four levels or major compositions. God as referred as *Latallah* resides at the very top (fourth floor) perceived as filled with supernatural joy and beauty over everything on earth, which is deemed to be a paradise. In this place, He occupied the highest hierarchy or level based on the belief that He is the most powerful supernatural being and creator of all creations on earth and this universe. *Latallah* does not depend on any other creatures, including humans to sustain or complete his existence, but can punish or bless humans and other creatures based on their behaviors. In this supernatural world, there also live angels and *malaikat*, but they occupy layers below the *Latallah*. The second level is reserved for the *embo'*, namely the spirits who have passed away long time ago and no longer depends on the conducts or actions of human life for their survival in the new world". "Not only because they had long passed away, but people or family members who were still alive when they left this world may have died and become a spirit or soul, and eventually became the occupant of the same realm. Although the *embo'* do not depend on the behaviour of the living, they are still respected, feared and remembered among those who are still alive because their title or status and aura still continue to dominate the life and are able to control the actions of living persons. This is because they are believed to always keep eyes on their children, grandchildren or great grandchildren and would advise them if they commit wrong words or wrong behaviours/conducts". "Thirdly, it is a level in which the world is divided into the real world and the supernatural world. As previously mentioned, the real world is inhabited by humans and other concrete objects or events, while the spiritual world is inhabited by supernatural creatures. In the supernatural world, there are particular residencies, namely the zones of air, land and water. In the air zone, there are two spaces which house the soul and spirit. Souls of the dead are placed in a column, while another column, which is above the soul space is allocated for the spirit. The soul will moves to the area of the spirit when the death is reaching 100 days".

3 *Tulah*

Derive from the world view mentioned above, the people of Bajau Samah also believe in the concept of *tulah* (condemnation), which is the belief that the catastrophe can occur among the people who violate the orders and commands of the King or Sultan and the supernatural. Stemming from animism[2], Bajau Samah believes that this world is not only inhabited by humans and animals, but also by the supernatural creatures that influence the lives of the humans.They believe that there are good creatures that have extraordinary powers in helping to solve the problems of

[2] Animism is the basis of religious belief that first emerged among the primitive man. Animism is the belief that every single thing on this earth has its spirit including the sun and the moon. They also believe that there are spirits and the supernatural who live with them. The animistic belief dictates that every single thing on this earth has its own soul that must be respected so that it will not interrupt humans, but will even help them in their daily affairs.

*Corresponding author: kipli@fmsp.upsi.edu.my

humans and there are evil creatures who give pain and adversaries in humans' lives (*tulah*). These creatures cannot be seen but their presence can be felt. Individuals who can see through these creatures are thought of as gifted and special, able to heal diseases caused by *tulah*. Individuals with this specialty will be regarded highly because they can help solve their problems – these people are called shaman or the witch doctors. Nonetheless, it cannot be spread to the public because it is feared that they will be disrupted by evil creatures. *Tulah* will happen when one commits evil deeds, or those who have violated the rules, for instance:

i. Disobeying parents and not showing any respect to other people
ii. Disturbing the areas of the supernatural.
iii. Disobeying the taboo of the custom.
iv. Hurt other people.
v. Talk badly about something without any valid proof.

In this vein, automatically the concept of *tulah* forms the societal organizational structure that always remembers and which does not commit anything bad. If one is plagued with *tulah*, he should see someone who is in good contact with the supernatural to heal his diseases, and this person is the shaman. All the rules and regulations, also taboo commanded by the shaman must be followed to avoid the *tulah* from being ongoing. The taboo that is always recommended by the shaman is the taboo concerning food and drinks in certain areas thought of the residential areas of the supernatural. "In this vein, the aspect of healing in *ngalai* is categorised as faith healing, which refers to the ritual practice that relates with the religious beliefs. Healing happens, in respect of the belief towards the spiritual power that has served as the intervention in the healing process" (Mohd kipli Abdul Rahman, 2015).

4 *Ngalai* Performance To Heal *Tulah*

From the belief on *tulah* there emerges the ritual performance of *ngalai* based on animism seeking to achieve healing. "It is a set of patterns from symbolic actions like certain forms of worship, sacrifice, and prohibitions. Rituals are the result of human desire in their pursuit of connection with the supernatural, as humans themselves would naturally possess some religious emotions. In the traditional context, performing arts was also a prevailing form of ritual used in the healing process. Ritual performance is a process that is created from the action or desire to connect with the metaphysical world as they tend to have some religious sentiments. It is also a medium for uniting the intention of the members of the society with a belief through various procedures and actions done symbolically like certain forms that originate from worship, sacrifice, and prohibition. Thus, ritual performance is a ceremony that comes from the action and desire to connect with the supernatural for certain purposes which are normally seeking to obtain good, wellbeing and peace within the practitioners" (Mohd Kipli Abdul Rahman, 2009).

The ritualistic performance often takes three days led by a shaman. According to Mohd Kipli Abdul Rahman, "in the context of the traditional Malay in specific, and the Archipelago in general, the role of the *bomoh* or shaman is as a mediator or connector between the physical world and the metaphysical world. This is because a shaman has his own strength, where he often has company, or is attached to the supernatural spirits that serve as an entity of the metaphysical world. Thus, a shaman can communicate with the entities from the metaphysical world in the effort to heal diseases believed to have been caused by the spirit" (Narupon Duangwises & Lowell D. Skar, 2016). This shaman is really respected by the people of Bajau Samah following his skills in healing various diseases using magic. This is because, Bajau Samah believes that *rojo*[3] (the supernatural creatures) who cures the patients indeed exists because during the process of healing, the shaman will communicate with *rojo*. *Rojo* determines the way that the patient is to be healed, and the things he must avoid throughout his healing period. The shaman in the ritual performance of *ngalai* often comprises of aging women.

For the purpose of performing for healing purposes, *sajian* (offerings) which comprise of some foods must be served for the supernatural creature (*rojo*) called upon to cure the patient. As coined by Mohd Kipli Abdul Rahman, "offerings (*sajian*) are vital in the healing ritual. The reason why *sajian* is prepared, is following the belief that there is a spirit (supernatural) that gives life to the ritualistic performance. Thus, when performing this ritualistic dance, the shaman has to ask for the permission first by calling upon the spirit or soul of the ancestors so that the performance will be blessed for healing purposes" (Narupon Duangwises & Lowell D. Skar, 2016). The foods come in the form of rice,

[3] *Rojo* (king) as the term used by Bajau Samah, is a supernatural creature that helps the shaman in the *ngalai* healing.

*Corresponding author: kipli@fmsp.upsi.edu.my

young coconut juice, glutinous rice, water and boiled eggs. Sometimes, female white chicken is also prepared as one of the foods served for the *rojo*. The coconut juice is prepared as drinks while the rice and boiled eggs are the foods for *rojo*. The foods will be added per request from *rojo* represented by the shaman. Meanwhile, incense is used to fumigate the performance and the patient areas.

When all is prepared, the *ngalai* performance will begin with the ceremony '*buka panggung*' (the opening of the hall) led by the shaman. The purpose is to steer clear from the disturbance of the evil. The ceremony begins by burning off the incense while fumigating it throughout the whole area and also around the patient and reciting some mantra. Several pieces of fabric were also used in the ceremony comprising of yellow and green clothes. Yellow cloth is used by the shaman to dab the body of the patient and dancers will use the yellow cloth to be wrapped around their waists. The green cloth is used by dancers and patient needs to cover his entire head. The reason or the dancers to wear yellow and green is to incite the spirit of the[4] *rojo* and his companions to exorcise the bodies of the dancers and the patient himself. The spirit is the handmaiden from the fairyland who accompanies *rojo* in the ritualistic performance of *ngalai*.

After the ceremony of 'buka panggung', the shaman will sing, followed by dancers dancing to the rhythm. The musical instrument used is a bamboo and a stick. The song is actually a mantra that calls upon *rojo* to come and bless the performance and fulfill the request made. Initially, the singing and the dancing follow a slow tempo. However, when it gets to the peak, it becomes increasingly fast and upbeat.

The singing and the dancing will stop when *rojo* and the shaman begin to interact through the patient's body that has been taken over by *rojo*. At the same time, the shaman will play his role in controlling the movements of the dancers who are also exorcised by the *rojo* handmaidens. Meanwhile, the shaman tries to regain the consciousness of the dancers from exorcism and lets them rest while he continues to communicate with *rojo*. During the process of communication, the shaman will wipe off the patient's face with a yellow cloth and dust off his body as a sign that *rojo* is present. In the communication, there are several things that the shaman will ask to the *rojo*, about the root cause of the disease he suffers, how to heal and the taboo. After all is asked, the shaman will recite again the mantra and prayers to extract *rojo* from the patient's body and send it back to the fairyland.

As it is sent home, the foods called '*pengeras*'[5] are also given as a sign of gratitude to the *rojo*. It is prepared as requested by *rojo*. The thing *rojo* normally asks is the white chicken known as *'ajung'*. *Ajung* will be swept away in the river right after the ceremony of *ngalai* for the healing to complete. After the ritualistic performance is over, the process of healing will occur and it will take a few days to heal entirely; it also depends on the strength of the patient's body. After the *ngalai* healing, the patient will feel extremely tired due to the exorcism he experiences. Normally, the family of the patient will hold a blessing ceremony or feast after the healing process so that the patient will heal even faster. The patient needs to adhere to several taboo determined by *rojo* mediated by the shaman to avoid from any unfortunate incidents or to avoid his disease from becoming ongoing. The healing of patients is seen as the physical and mental capabilities to manage himself or herself as normal.

5 *Ngalai* Ritual In Today's Context

The ritualistic performance of *ngalai* for healing purposes serves as a method that gives a great influence to patients especially diseases caused by the disturbance of the supernatural (*tulah*). In this matter, patients have full reliance on the shaman and the public to heal themselves with the help of *rojo*. Therefore, the process of healing through *ngalai* is regarded as *communal healing* based on the method adopted involving the patients and the public. It is a dynamic process done through the collective experiences and experimentation. The main objective of the communal healing process manifested through the ritual of *ngalai* is to revive the balance between the physical and the metaphysical through the healing of patients caused by *tulah*. In relation to that, the "aspect of healing in *ngalai* is also categorised as faith healing, as healing happens in respect of the belief towards the spiritual power that has served as the intervention in the healing process".

---

[4] The soul that exists in all the creatures (things) either dead or alive and which traditionally was believed to be able to give strength or to become evil

[5] **Pengeras** – is a token of appreciation presented by the patient to the shaman as requested by the *embo'-embo'* in a form of foods or sometimes money.

*Corresponding author: kipli@fmsp.upsi.edu.my

Thus, in the current context of the people of Bajau Samah who still believe in animism, they also believe in the concept of *tulah*. Among the people of Bajau Samah there are still individuals known as *bomoh* or shaman, who have the knowledge to heal diseases inflicted by *tulah*. The diseases caused by *tulah* cannot be detected through modern medicine as it relates to the supernatural from the metaphysical universe. The evidence that *ngalai* is still significant in the context of Bajau Samah is through the practice of the ritualistic performance of *ngalai* itself which still exists and which is still practised by the people.

6 Conclusion

In terms of the performance, *ngalai* is a combination of elements of dance movements, singing, music and acting. Traditionally, the ritual of *ngalai* is a performance practice to heal, based on the animistic religion that believes in the existence of the supernatural. Individuals who can see these creatures are said to be very able and they have the specialty in healing diseases caused by *tulah* following the disturbance of the supernatural. The healing aspect in *ngalai* is categorized as communal healing as well "as faith healing", referring "to the ritual practice that relates to religious beliefs and healing" that happen following the method adopted involving patients and the public. It is a dynamic process done through the collective experiences and experimentation. This study finds that the ritual of *ngalai* is a ritualistic artistic performance that functions as a medium of healing and it is still significant in the context today.

Acknowledgments

The authors acknowledged "Universiti Pendidikan Sultan Idris, Malaysia" and "the Ministry of Higher Education Malaysia for the funding of this research under the Fundamental Research Grant Scheme" (FRGS) 2016-0077-107-02 (FRGS/1/2016/WAB04/UPSI/02/2) and "Long Term Research Grant Scheme (LRGS), Ministry of Education Malaysia, 203/PTS/6720007, 203/PTS/6727001, 203/PTS/6727002, 203/PTS/6727003". Our appreciation also goes to Bajau/Sama' community in Kota Marudu, Kota Belud and Semporna, Sabah, for providing the prime information assistance.

References

1. "Gang Chen, M.A. (2000). Death Rituals in a Chinese Village: An Old Tradition in Contemporary Social Context. Unpublished PhD dissertation, The Ohio State University".
2. "Geertz. C. (2008). Tafsiran Budaya/The Interpretation of Cultures. (Trans.). Kuala Lumpur: Institut Terjemahan Negara Malaysia Berhad. (Original work published 1973)".
3. Halina Sendera Mohd. Yakin. (2013). "Cosmology and World-View among the Bajau: The Supernatural Beliefs and Cultural Evolution. *Mediterranean Journal of Social Sciences.* Vol 4" No 9.
4. "Mohd Kipli Abdul Rahman. (2009). *Mabuk mistikal: Semiotik metafizik dalam kuda kepang mabuk*. Penang: USM Publisher".
5. "Mohd Kipli Abdul Rahman. (2015). Bebayuh: From healing ritual to traditional performance. *IJASOS - International E-Journal Of Advances In Social Sciences*, 1(2), 136-141. doi:http://dx.doi.org/10.18769/ijasos.75810".
6. Narupon Duangwises & Lowell D. Skar (Editor). (2016). *The Folk Performing Arts in ASEAN*. Bangkok: Princess Maha Chakri Sirindhorn Antropology Centre, SAC Publication.
7. "Juli Edo. (2001). Kosmologi Orang Asli: Satu Pemerihalan Tentang Perjalanan Roh. In Yaacob Harun (Ed.), Kosmologi Melayu (pp. 133-146), Kuala Lumpur: Akademi Pengajian Melayu, Universiti Malaya".
8. "Singaravelu, S. (2001). Kesan kosmogini dan kosmografi India di alam Melayu. (Yaacob Harun Ed.). Kosmologi Melayu. (pp. 43-51), Kuala Lumpur: Akademi Pengajian Melayu, Universiti Malaya".

*Corresponding author: kipli@fmsp.upsi.edu.my

Authentication of a selected medicinal plants using DNA barcoding technique

Che Nasriyyah Che Husin[1,] *Darlina* Md. Naim[1] , *Mardiana Idayu* Ahmad[2]
[1]School of Biological Sciences, Universiti Sains Malaysia, 11800 Pulau Pinang, Malaysia
[2] School of Industrial Technology, Universiti Sains Malaysia, 11800 Pulau Pinang, Malaysia

Abstract. Plants are valuable source of a medicine and have long being used to cure various ailments. However, the efficacy of drugs derived from plant depends on the reliable identification of correct plants. To avoids the usage of incorrect plant that can cause poisoning, a reliable method than morphological characteristic is required. DNA barcoding technique have shown to be an efficient tool for species identification by using a short fragment of the genomic DNA and has been used widely in molecular plant taxonomy for authentication of medicinal plants species. Thus, the goal of this study was to use DNA barcoding technique to discriminate medicinal plants. DNA samples were extracted from twenty medicinal plants, chosen based on their therapeutic efficacy and were used as templates. Internal transcribed spacer (ITS2) gene was selected to be the best molecular marker for identification purposes. The efficiency of the amplification by polymerase chain reaction was sending for sequencing and species identification was performed using MEGA6. Our findings show that DNA barcoding is an efficient tool for plants identification. This study revealed that medicinal plant and their closely related species can be distinguished by using DNA barcoding technique with ITS2 region as it is an efficient marker and potential DNA marker for authentication of selected plants.

1 Introduction

Plants have been used as source of medicine for thousands of years by many cultures and there are many publications [1, 6, 9, 10] that highlighted the usage of plants to treat various kinds of diseases. In recent years, traditional herbal medicines have gained popularity as consumers are reportedly more comfortable taking product that are deemed natural, easily available, cheap and are less side effects. Approximately about one-third of commercially drugs in pharmaceutical industry are natural products based origin or their derivatives.

The traditional approach towards discovery of plant based drugs often involved amount of time and expenditure. Traditional approaches for identification of herbal products include organoleptic methods (identification by sense: taste, sight, smell, touch), macroscopic and microscopic methods (identification by shape, colour, texture) [13]. However, in most cases a professional taxonomist is needed. If the specimen is damaged or immature stage of development, even specialist may be unable to identify. Another identification method is by using chemical profiling but it may be affected by physiological and storage conditions. Therefore, there is a need for new approach in an efficient and accurate identification of plant species and their adulterants. Authentication using DNA barcoding overcomes these problems.

DNA barcode is a rapid and accurate identification of any plant species based on extracting a short DNA sequence from a tiny sample of any organism. It allows taxonomists to quickly and accurately recognize species and retrieve information about them. It also allows non-expert to identify species from damaged or industrially processed material. The term DNA barcode for species identification was introduced by Paul Hebert in 2003 (Hebert & Gregory, 2005). Several potential existing barcoding candidates have been reported such as *atpF-atpH*, *matK*, *rbcL*, *psbK-psbI*, *trnH-psbA*, *ycf5* and *ITS* [7, 11].

*Corresponding author: darlinamdn@usm.my

2 Materials and methods

2.1 Plant material

In this study twenty matured fresh leaves of selected medicinal plants were collected. Damaged leaves were sorted out followed by washing with running water to remove the dirt and adherent.

2.2 DNA preparation, amplification and sequencing

DNA was extracted from leaves using modified CTAB method [2]. DNA were amplified by polymerase chain reaction (PCR) through 30 cycles of 95 °C for 2 min, 94 °C for 45 s, 47.9 °C for 45 s and 72 °C for 1 min. The primers used were *ITS2-F*, 5'-GCGATACTTGGTGTGAAT-3' and *ITS2-R*, 5'-GACGCTTCTCCAGACTACAAT-3' .The PCR product were run on a 2% agarose gel in 0.5 x TBE buffer and DNA sequencing was performed by Bioneer Corporation located at Korea using the same amplification primers.

2.3 Sequence analysis

Sequencing results was aligned using MEGA6 software [12]. The sequences of the species were blasted with the GenBank database to assess the identification of unknown sequence. The highest-scoring hit of maximum identity from each sample was taken as the barcoding identification. Results were given in Table 1.

3 Results and discussion

DNA barcoding have been applied in the field of taxonomy and become popular in the past few years [7]. The technique is a useful tool for taxonomist because rapid and accurate method of species identification is very much important to ensure that the drugs made from medicinal herbs are safe to consume. There is an urgent need to establish a rapid and simple DNA barcoding identification system, since scientist in other field also a potential user of DNA barcodes [14].

Current candidate barcodes are ITS and ITS2 of nuclear ribosomal DNA, as well as psbA-trnH, matK, ycf5, rbcL and rpoC1 in the chloroplast. Previous observations have been shown the potential of ITS2 as a suitable marker for taxonomic classification compared to other barcode markers [15]. In this study, we used internal transcribed spacers 2 (ITS2) as DNA barcodes to identify selected medicinal plants. The ability of routine amplification using a universal primer pair and the generation of unambiguous bidirectional DNA sequences with minimum editing are two major criteria of selecting an ideal DNA barcode. In addition, ITS2 has good primer sites due to the region is short which is easy to amplify and sequence [17].

Results from the study demonstrated that the primer used, ITS2 presents a promising universal DNA barcode to recognize twenty selected medicinal plants by giving high percentage of maximum identity for each sample (Table. 1). However, a few samples showed low percentage of maximum identity below than 95% which are *Pereskia sacharosa*, *Murraya koenigi* (84%), *Piper sarmentosum* (88%), *Carica papaya* (87%) *and Clinacanthus nutans* (87%). Maximum identity below than 95% was rejected because the result shows an ambiguous species identification [5].

Table 1. Sequencing result

No	Scientific Name	Vernacular Name	Maximum Identity (%)
1	*Emilia sonchifolia*	Bayam paksi	98
2	*Kaempferia galanga*	Cekur	98
3	*Brucea javanica*	Lada pahit	99
4	*Pereskia sacharosa*	Duri 7	84
5	*Peperomia pellucida*	Ketumpang air	99
6	*Citrus maxima*	Limau bali	100
7	*Alpinia galanga*	Lengkuas	99
8	*Mangifera indica*	Mangga	99
9	*Amaranthus tricolor*	Bayam	100

10	*Sauropus androgynous*	Pucuk manis	99
11	*Oenanthe javanica*	Selom	99
12	*Murraya koenigii*	Kari leave	89
13	*Averrhoa bilimbi*	Belimbing buluh	97
14	*Chromalaena odorata*	Kapal terbang	100
15	*Allium tuberosum*	Kucai	99
16	*Psidium guajava*	Jambu batu	100
17	*Piper sarmentosum*	Kaduk	88
18	*Carica papaya*	Betik	87
19	*Clinacanthus nutans*	Belalai gajah	87
20	*Melastoma imbricatum*	Keduduk puteh	100

Maximum identity is an indicator that used to measure the likeliness of identified species. If the maximum identity more than 95%, it is likely to have been correctly identified. Nevertheless, if the closely related species to the identified species in geographical area were not included in the reference database, .even the maximum identity more than 95%, the identity may be incorrect.

The successful of species identification influenced by some factors. First, the quality of yield DNA, the isolation of pure and high molecular weight DNA is critical for the successful application of molecular methods and this can be quite challenge since the DNA is often highly degraded by high temperature. Second, a suitable primer used and third, the sequence quality [8]. The successful rates of DNA sequencing can be relatively low due to the presence of polymorphic DNA in multiple copies [16] and also because the sequences were contaminated by fungal or other parasitic species, such as arthropods and protozoa [4].

4 Conclusions

DNA barcoding is a reliable tool for authentication of the selected medicinal plants. In this study, the DNA barcoding with the use of ITS2 from the nucleus gene was highly successful. Therefore, our finding contributed to the previous studies of ITS2 as a potential marker that can be implemented for studying and identifying medicinal plants.

5 Acknowledgements

This study was supported by Long Term Research Grant (LRGS) on 'Conceptualizing local knowledge: A study on sustainable values of local wisdom and indigenous science' (R/LRGS/A02.00/00559A/004/2012/000089).

6 References

1. W.W. Chao, B.F. Lin, Chinese Medicine, **5(1)**, 1 (2010)
2. J.H. Cota-Sánchez, K. Remarchuk, K. Ubayasena, *Plant Molecular Biology Reporter*, **24(2)**, 161-167 (2006)
3. P.D. Hebert, T. R. Gregory, Syst Biol, **54(5)**, 852-859 (2005)
4. P. M. Hollingsworth, S. W. Graham, D. P. Little, PLoS One, **6(5)**, e19254 (2011)
5. A. Kool, H. J. De Boer, A. Kruger, A. Rydberg, A. A. Abbad, L. Bjork, G. Martin,PLos One, **7**, e39459 (2012)
6. S. Y. Lin, C. C. Wang, Y. L. Lu, W. C. Wu, W. C Hou, Food and Chemical Toxicology, **46(7)**, 2485-2492 (2008)
7. K. Luo, S. Chen, K. Chen, J. Song, H. Yao, X. Ma, X. Li, Science China Life Sciences, **53(6)**, 701-708 (2010)
8. P. Mahadani, S. K. Ghosh, DNA Barcodes, **1**, 35-38 (2013)
9. T. Nalina, Z. Rahim, Am J Biotechnol Biochem, **3(1)**, 10-15 (2007)
10. M. J. Oh, M. A. Hamid, S. Ngadiran, Y. K. Seo, M. R. Sarmidi, C. S. Park, *Archives of dermatologicalresearch*, **303(3)**, 161-170 (2011)
11. X. Pang, J. Song, Y. Zhu, C. Xie, S. Chen, Planta medica, **76(15)**, 1784-1786 (2010)
12. K. Tamura, G. Stecher, D. Peterson, A. Filipski, S. Kumar, *Mol Biol Evol*, **30(12)**, 2725-2729 (2013)
13. N. Techen, I. Parveen, Z. Pan, I. A. Khan, *Current Opinion in Biotechnology*, **25**, 103-110 (2014)
14. M. W. Chase, N. Salamin, M. wilkinson, J. M. Dunwell, R. P. Kesanakurthi, *Philosophil Transactions of The Royal Society B*, **360**, 1889-1895 (2005)

15. A. W. Coleman, Trends in Genetics, **19**, 370-375
16. I. Alvarez, J. F. Wendel, *Molecular Phylogenetics and Evolution*, **29(3)**, 417-434 (2003)
17. W. J. Kress, K. J. Wurdack, E. A. Zimmer, L. A. Weight, D. H. Janzen, *The National Academy of Sciences*, **102**, 8369 (2005)

Knowledge transfer and management among Malay traditional massage practitioners in the north of Malaysia

Norizan Esa [1,1], *Salasiah* Che Lah[2], and *Sakinah* Mahamad Hakimi[3]

[1]School of Educational Studies, Universiti Sains Malaysia, 11800 Pulau Pinang, Malaysia
[2]School of Humanities, Universiti Sains Malaysia, 11800, Pulau Pinang, Malaysia
[3]Freelance researcher, Pulau Pinang, Malaysia

Abstract. Local knowledge is knowledge transferred across generations by individuals or community through experience in the environment they live in. This includes the knowledge and practices that are still practiced by a community or individual. Apart from trust, the willingness and capacity to share what they know and use what they learn plays an important role for local knowledge to be transferred and managed. In Malaysia, traditional Malay massage is a form of traditional healing among the Malay society that was recorded since the 19th century through *Kitab Tib* manuscript. This form of healing uses natural resources, *wafak* (letters, numerals and diagrams), Quranic verses and *doa* (prayers and supplications) in the healing practice. This paper attempts to explore on how the knowledge is passed down from expert to apprentice. This study involves interviews and observation on selected Malay traditional massage practitioners. In conclusion, knowledge transfer among Malay traditional massage practitioners only happens when there is trust trust. Knowledge is passed down within their family members and learned from observation and experience. The traditional Malay massage remains only within the individual or a limited number within the family community rather than the community as a whole.

1 Introduction

Local knowledge is a knowledge passed on through experience from one generation to the next generation by individual or a community within the environment they live in. Local knowledge includes the knowledge, practices that are still practiced by a community or individual including language, place, spirits and world view [1]. Generally, the knowledge is transferred and managed when an individual or community trust each other and gain benefit from the transfer process.

Previous study found that, 'tacit knowledge' is not easy to be translated into written form such as documents, blueprints, and procedures. This knowledge can only be acquired from personal experience and difficult to be transferred and formalized. The transfer of knowledge usually learnt through apprenticeship via shared and collaborative experiences from expert to novice. The learning process requires 'doing' and participation from an expert to apprentice requires 'trust' for one to obtain the knowledge and skill desired. The higher the trust between the expert and novice, the more successful for the knowledge to be transferred [2]. For example, the transmission of environmental knowledge and land skills of 47 traditional Inuit men in Ulukhaktok, Canada. A list of 83 skills on the importance of safe and successful harvesting was passed down and transferred from older to younger generations through observation and apprenticeship in the environment. The loss of certain skills on hunting, fishing, trapping and incomplete transmission of knowledge was due to lack of skilled teachers, loss of native language, and changes in the educational environment [3].

The transmission of knowledge can also take place when there is a shared goal for technology and knowledge to be exchanged. The persistency of a novice in learning the knowledge is also an important factor for a knowledge to

[1] Corresponding author: salasiah@usm.my

be successfully transferred. Other than that, technical capability, strategic and cultural compatibility also becomes one of the success factors for the knowledge to be transferred. The more persistent the novice show in learning the knowledge, the higher the trust between the expert and novice. Through persistence and trust, the knowledge can be exchanged easier [2].

An example of tacit knowledge in Malaysia is the knowledge on traditional massage, whereby it falls under Traditional and Complementary Medicine. Traditional massage is practiced together other than the practice of medicine or surgery by registered medical practitioners defined in the Medical Act 1971 approved by Malaysian Medical Council, Ministry of Health Malaysia [4]. In the context of local knowledge, the knowledge is learned from personal experience, shared and collaborative experiences which require participation and 'doing'. However, since local knowledge can only be obtained through trust and shared goals, the transfer of local knowledge only take place once these two factors shown by both the expert and novice which is the apprentice chosen by the expert of particular knowledge. The expert and apprentice will only exchange confidential information, after they have assessed the competence and reliability of the person who will receive the knowledge. The identified apprentice will be assessed by their technical capabilities, strategic and cultural compatibility. They will also negotiate the terms of the partnership depending on the participation, length, breadth and the way the knowledge is communicated. Due to this, local knowledge usually shared and passed on between trusted family members and rarely to the community. Instead it is passed on within family community such as from parents to their children, from grandparents to their grandchildren or from uncle or aunts to their niece or nephews. The expert in the family such as grandmother or the parents will identify and carefully select their apprentice to learn and practice the knowledge they have. However, not all knowledge will be shared and transferred to their apprentice due to their own reason to protect the knowledge.

According to World Health Organization, traditional healing has been defined as a diversity of practices, approaches, knowledge and beliefs incorporating plant, animal and mineral-based medicines. This includes, spiritual therapy, techniques, and exercises applied individually or combined to maintain health, treat illness or prevent it. Traditional healing has been widely used as prevention, diagnosis and treatment of an extensive ailment. It is an important alternative form of healthcare service to a person or community who is with or without geographic and financial access to modern medicine [5]. The Ministry of Health Malaysia defined traditional healing as traditional and complementary medicine other than practice of medicine or surgery by registered medical practitioners as defined by Medical Act 197. There are five traditional complementary medicine practiced in Malaysia which is Malay, Chinese and Indian traditional medicine, Homeopathy and Complementary Medicine, including traditional Malay massage [4].

In Malaysia, traditional healing among the Malay society was recorded since the 19th century through *Kitab Tib* manuscript which uses natural resources, *wafak* (letters, numerals and diagrams), Quranic verse and doa (prayers and supplications) during the healing practice. At times, traditional Malay massage practitioner uses traditional healing as part of their practice whereby it uses a diversity of practices, approaches, knowledge and beliefs incorporating plants, animal, and mineral based medicine. This may include spiritual therapies, techniques and exercises, applied individually or in combination to maintain health in treating illness or to prevent it.

In the Malay society, healing practices utilizes natural resources mainly flora as remedies because it gives benefits for the body aligned with Islamic teachings. In an analysis on *'Kitab Tib'* alone, the author recorded about 258 types of flora that can be found mostly locally. The use of herbs as the ingredients in these remedies proves that the Malay society already have its own traditional healing practices that was passed down from generation to generation even until today [6]. By participating, the knowledge on tacit knowledge can be transferred through learning and living practice between expert and novice. Learning is also seen as a way to gain respect from the community whereby skills are developed and incorporated through training and practice that can provide tangible benefits to the community [3].

The willingness and capacity of individuals or a community to share what they know and use what they learn plays an important role for local knowledge to be transferred. Therefore face-to-face interaction becomes the primary means in obtaining local knowledge from the source of knowledge [7]. Since that local knowledge are learnt from 'doing' and participation between the source and receiver of knowledge, local knowledge are more successful to be passed down if a face-to-face interaction takes place whereby the knowledge can be better exchanged and new knowledge that can improve the current practices of the identified local knowledge be developed. Expertise is developed through learning-by-doing, an apprentice are likely to practice under the guidance of a knowledgeable teacher for them to become an expert in the field. The more engaged a practitioner is with the activity, the better they become good at it

[8]. Therefore, this paper attempts to explore on whether or not these traditional Malay massage practitioner would transfer their knowledge and how the knowledge is being passed down from expert to apprentice.

2 Methodology

This study involves selected traditional Malay massage practitioners in the northern of Malaysia on how they acquire their knowledge on traditional massage and how they manage the knowledge obtained. It also attempts to observe whether or not these traditional Malay massage practitioners would transfer their knowledge to their apprentice. Traditional massage practitioner is an expert in massaging people traditionally as their main activity. However, some practitioners may also offer massage treatment on women who recently gave birth. There were seven people involved in this study but for the purpose of this research, only four will be discussed.

3 Results and Discussion

According to Respondent A, a 34 year old traditional Malay massage practitioner, he learns how to massage people through observation from his grandfather who is also a massage practitioner. The knowledge usually would be passed down when an apprentice is in their 40s but because of his grandfather's health is deteriorating he had to learn from his grandfather as early as 18 years old. He doesn't know why out of all 80 grandchildren, he was chosen by his grandfather to learn about traditional massage. Upon returning to his village after being informed about his grandfather's health, he learns how to massage people from his grandfather after Maghrib prayer for a day. His grandfather use verses from the Quran in his healing practice as he believes that all diseases happen by the will of God. Method of healing depends on the type of disease of the patient. Other than that he also learns the type of food that needs to be avoided or eaten during healing process.

This means, out of all his grandchildren, Respondent A grandfather have already assessed who among them who is suitable to be passed down his knowledge on massage. Hence why, he was chosen as an apprentice and transferred his knowledge from his grandfather.

Through his grandfather, he not only learns how to heal but also learns to be a *'bidan'* or midwife. Although he has the knowledge as a *'bidan'*, he doesn't deliver baby because he is a man and cannot touch women's private part. In spite of that, he learns how to make *'Minyak Celak'* and *'Minyak Susuk'* used by women. However, he usually buys the *'Minyak Urut'* or oil for massage as the ingredients can only be obtained from the forest. As many people started to be convinced of his ability, his ability to heal people spread through word of mouth and receives patients from people in the south of Malaysia. Sometimes, he also receives non-muslim patients who come for massage after giving birth. During massage, he takes about one to two hours depending on the individual and the size of their veins. Every month, he treats more than 10 patients and receives payment in whatever his patients can afford. Although he doesn't set specific time to treat his patients, he does other work such as farming.

From this, we observe that his ability to massage people manage to gain the trust by the people who came for his treatment. People are more convinced as they hear from a other people they trust and seeks his treatment for massage as word of mouth is a powerful tool and easily believed as it is built on trust for a person to accept the information they get.

Respondent A also adds on that although he knows how to heal people such as *'Saka'* , hysteria, possessed, he do not know particular type of massage technique for broken bones because he is still young. His grandfather have set a criteria that only apprentice who reach the age 38 years old are qualified to learn how to treat broken bones. Therefore, he also continues the same method to people who wants to learn from him. He hopes to pass down his knowledge to his children when they reach the right age to learn which is between 30 to 40 years old. However, he is willing to teach other people within his family members depending on how much they can learn and compatibility of the apprentice in learning. For example, he tried to pass down the knowledge to his brother in law but after 44 days of learning one method, his body heat increased for a week and cannot withstand it. So they had to remove it and the transfer was not successful.

From the interview, we learn that Respondent A grandfather sets a certain criteria in order to select the most suitable apprentice and he tries to continue the method by passing down to his brother in law. According to his grandfather way, an apprentice should reach the age of 38 years old to be qualified to learn about massaging and treating broken bones. However, the knowledge could not be transferred as the brother in law couldn't stand the increasing body heat after a week of their apprenticeship. Although he is willing to teach other people within his family members, he still hopes that after his children reach the right age to transfer his knowledge, they will be able to learn and continue the knowledge passed down to them.

On contrary, for Respondent B, she learns from her great grandmother from her father side who is also a *'Bidan'*. None of her siblings knows how to massage except her. She start giving service when she is 40 years old. Similarly with Respondent A, she doesn't know how to treat broken bones and only knows how to massage after giving birth and deliver baby. However, Respondent C makes her own massage oil from coconut and uses verses from the Quran in her healing practices. Normally, she starts massaging after 3 days when her patients gives birth and offers *'cukur jambul'* service to shave a baby's hair. Her ability was spread through word of mouth and at times was invited to treat people in other cities such as Langkawi, Pulau Pinang, and Klang. Every massage also takes about one or two hour the most and she will advice on the type of food to avoid or eat during the healing process. However, some practices she no longer continued by new mothers as it is against the teachings of Islam. Some of her children have the ability to massage so other people will hire them as she is getting old. Although she practices traditional massage, Respondent B also uses modern complimentary medicine such as *Gamat* gel to heal wounds and goes to courses for midwife. However, she doesn't teach anyone her knowledge because no one wants to learn from her.

Meanwhile, for Respondent C the knowledge on massage was passed down quite mysteriously. She learns it through dreams and receives a black mark in her palate as a sign of *'wali'* or a respectable person who have special power which she believes inherited in her family. Because of the black palate, her family is known as *'Tanah Hitam'* or the black soil family. Similarly, she also use verses from the Quran in her healing practice. It is believed that whatever her family says will become true when it is said. Therefore, she uses her ability to make *doa* and prayer for the person who comes for treatment with very low fee. Strangely, she charges only RM1.20 for her service which is optional. Her reason was that she only wanted to help people with her ability

In Respondent D case, she inherited naturally from her family who is also a *'bidan'* or midwife. She have the ability treat broken bones and sprained parts of the body. But for those who still cannot recover and asks for spiritual healing, she will politely decline. Since her mother was also a *'bidan'*, she also inherited the ability from her mother and the healing knowledge from her father's side who inherits a spiritual tiger passed down to their family to protect them from harm. Despite that, she uses Quranic verse in her treatment and certain type of Islamic prayers for treatment. However, she only treats women during massage and only massage up to the feet if she had to massage male patients. Her patients are required to bring a coconut which she will make oil from it where they will later use for massage. Sometimes, the patient brings their own oil massage which she later make *doa* and blow on the oil massage they brought. Usually she starts her service around 2 to 6 pm and only gives service when asked to help. Some patients come all the way from Johor and Kuala Lumpur. According to Respondent D, none of her children have the ability and talent that she have now. Her youngest children who showed ability like her have passed away which she believes is because her child might grow up to become a bad person. She believes that those with her ability cannot talk about bad things and be mad at people. Although her grandchildren seem to show ability like her, she is not sure if she can passed down the knowledge as it is passed down through dreams like her.

Conclusion

To sum, knowledge transfer among traditional Malay massage practitioners can only take place when there is trust between the expert and apprentice. This knowledge is often passed down within their family members only. The skills and technique on traditional massage is learnt by the apprentice from experience and observation of the expert. However, some local knowledge cannot be passed down either because the apprentice is not ready in many forms (i.e. spiritual strength, compatibility, age maturity) to receive the knowledge or nobody is willing to learn from these experts and naturally gifted within particular family. Some practitioners don't necessarily stick to the same method and may generate their own knowledge in treating their patients by mixing with other types of massage practices. Although there is modern technique for massage available, rather than learnt by the community as a whole. The traditional massage remains only within the individual or limited number within the family community.

Acknowledgement

The authors express their gratitude to the Malaysian Ministry of Higher Education for awarding the Long-Term Research Grant (LRGS, 203/PTS/6727002), Universiti Sains Malaysia and all involved in this study

References

1. Canagarajah, S. (2002) Reconstructing Local Knowledge. *Journal of Language, Identity & Education*, 1:4, 243 – 259.
2. Rothernberg et. al, (2006) Tacit knowledge transfer and the knowledge disconnect. *Journal of Knowledge Management*, Vol. 10 pp. 6-18.
3. Furgal et. al (2011) "Transmission of environmental knowledge and land skills among Inuit Men in Ulukhaktok, Northwest Territories, Canada", Human Ecology Vol. 39, Issue 3, pp 271 – 288.
4. Pillay, M. S (2006) "Traditional & Complementary Medicine in Malaysia", Ministry of Health Malaysia, International Conclave on Traditional Medicine, New Delhi, India. 16-17
5. (WHO, 2001).
6. Baharuddin & Sidek (2010) Traditional healing in Malay culture: The Case of a Malay Manuscript of the 19th century , International Conference of Sustainable Health Promotion.
7. Wang & Wang (2012) Knowledge sharing, innovation and performance, *Elsevier*, Expert system with applications.
8. Abrams. L, Shields. M, Leonard. D et al. (2015), Using mentoring and storytelling to transfer knowledge in the workplace, *Routledge*, Journal of Management and Information Systems

Adaptability to Settlement Pattern and Choice of Subsistence Activities: Emergence of Material Culture within the Saribas Malay in Betong, Sarawak

Faridah Sahari[1*], *Anna* Durin[1], *Rahah* Hasan[1], *Shahren* Ahmad Zaidi Adruce[2] and *Shahri* Abdul Rahman[2]

[1]Faculty of Applied and Creative Art, Universiti Malaysia Sarawak
[2]Faculty of Cognitive Sciences & Human Development, Universiti Malaysia Sarawak

Abstract. Despite many job opportunities in the market and the challenges they have to face, some minority of the Saribas Malay community in Betong, Sarawak is still maintaining traditional subsistence activities in food production based on the *nipah* tree (locally known as *apong*) such as *gula apong, garam apong cuka apong, jarik mayang, air sadap* and the sago tree (locally known as *mulong*) produce, *lemantak*. This research examines the choice of subsistence strategies and settlement pattern of the Malay community who inhabit the Saribas region. Through the in-depth interview and participant observation, the finding suggested that reliance on a river as the main highway to connect them to the other parts of Sarawak and river as a source of marine resources determine the choice of linear settlement pattern along the river. The results also suggest that river terrestrial resources; *apong* and *mulong* accessibility and abundance availability that influence the community in continuing the traditional subsistence activities (*apong* and *mulong* based food production) related to those flora source. As such, the assemblage of material culture that exist within the Saribas Malay community is the representation and manifestation of their choice of settlement pattern and subsistence activities.

1 Introduction : The Saribas Malay community

The Saribas state has long existed. According to Sanib (2012, 2013), the Saribas, was among the other four Lost Kingdoms toponyms; Sarawak, Samarahan, Kalaka and Melanau, that appeared in the Brunei Sultanate various manuscripts and in the Javanese chronicle ever since 1365. Due to restricted sources of the Saribas Malay available in the present, this paper is an early attempts to documenting the material culture of the people of once been the prominent state in the Sarawak history. Three aspects of Saribas Malay are discussed in this paper: settlement pattern, livelihood strategy and material culture that emerge from the sustenance activities of the community; which is based on the premise that these three aspects are interrelated and characterized by their environment.

Cultural diversity of a community is affected by various factors associated with the customs, traditions, ways and lifestyle of the community. The objects of the Sarawak Malay culture are closely related to socioeconomic activities, and each one has its function and role, and vary significantly with the Malay in Peninsular Malaysia (Awang Azman, 2014). In fact, the term (in Sarawak dialect) used for the cultural objects is also unique among the Malay community from different parts of Sarawak (Mary Fatimah and Salbia, 2014).

* Corresponding author: sfaridah@unimas.my

The Malays constitute the majority of the people in Betong Division. According to data released in 2013 by the Sarawak Planning Unit (SPU), the two colonies in the division administration, Betong and Saratok cover an area of 4181km square and 59,000 (52%) of the total 114,000 population is Malay.

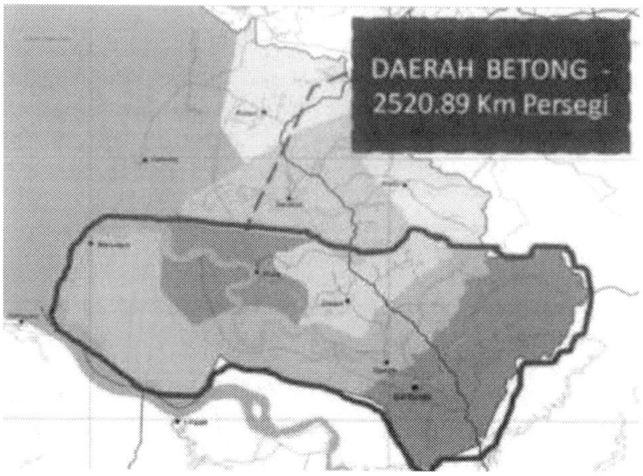

Figure 1 Betong District Map (Source: Betong District Council, 2017)

Table 1 shows the population by ethnic group of five sub-districts of Betong, and Malay appeared to be the largest population in Pusa. This paper is about the Saribas Malay of Tambak, Pusa in Betong Divison Sarawak. The location is selected based on the active *apong* and *mulong* related activity in the area.

Table 1 Total Population by ethnic group, sub-district and state, Malaysia, 2010
(Source: Department of Statistics Malaysia)

Daerah Pentadbiran/ Daerah kecil *Administrative District/ Sub-district*	Jumlah *Total*	Warganegara Malaysia *Malaysian citizens*										Bukan Warganegara Malaysia *Non-Malaysian citizens*
		Jumlah *Total*	Bumiputera						Cina *Chinese*	India *Indians*	Lain-lain *Others*	
			Jumlah *Total*	Melayu *Malay*	Iban	Bidayuh	Melanau	Bumiputera Lain *Other Bumiputera*				
BETONG	60,728	60,216	57,730	35,917	21,293	195	148	177	2,371	34	81	512
Betong	17,014	16,955	15,551	2,789	12,482	116	60	94	1,319	20	65	59
Debak	10,276	10,206	9,789	4,447	5,262	33	12	35	401	5	11	70
Maludam	5,901	5,826	5,717	4,901	769	11	15	21	106	1	2	75
Pusa	14,921	14,626	14,422	14,231	118	21	38	16	198	5	1	295
Spaoh	12,616	12,603	12,251	9,549	2,652	14	25	11	347	3	2	13

2 The relationship between settlement pattern, environment and subsistence activity

2.1 Settlement pattern and environment

The Saribas Malay was named after a name of a river, which is Batang (local term for river) Saribas. The Malay community was settled near the river bank and along the coastal line of Saribas River and since the mid of the 16th century, they barter traded the salt and salted fish with the Dayak for rice and jungle produce (Sabihah, 1972). The barter trade materials (salt: presumably *garam apong* and salted fish) mentioned in Sabihah's article suggest that, ever since the 16th century, the Saribas Malay have relied on the river and its terrestrial resources in their food production technology and subsistence activities.

Figure 2 shows the distribution of major forest types in Sarawak, indicates the intensified of peat swamp forest in the region of Betong district. The geomorphology of this type of forest is on low, swampy and muddy land near to brackish river water terrestrial. Here is where *apong* (*nipah* palm) and *mulong* (sago palm) grow, among other diverse typical swamp vegetation of river mangrove and peat swamp (Johnson, 1992; Tsuji K. *et. al.*, 2011). That explains the abundance of these two plants, fringed along the riverside of the Saribas where the Malay village settlement area is situated.

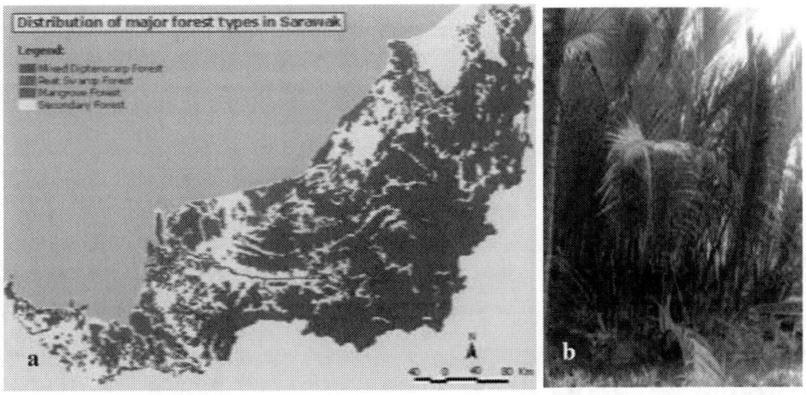

Figure 2 a) Forest Types in Sarawak (*Source: Sarawak Forestry Corporation, 2017*) and b) *nipah* palm tree

Despite wage labour employment and the scarcity of *apong*, economically, some minority of the Saribas Malay community in Kampung Tambak are still relying on that the palm tree as a main source of their subsistence activity; *apong* related food production. For the purpose of this study, the respondents are purposively selected based on two criteria: 1) their extended (15 to 20 years) of experience in *gula apong* production and 2) their families have been involved in in *gula apong* production for many generations. According to the respondents; Yahya, Rosman and Rosnah, their family had been relying on this activity a means of earning a living. *Gula apong* (palm sugar), *garam apong* (palm salt), *cuka apong* (palm vinegar) and *lemantak* (sago) are among the common foods related to *apong* and *mulong* produced by the community in Tambak, Pusa.
The researchers have been visiting the location of study for several times to interview the respective respondents and observe *gula apong* manufacturing process. Documentation is done through video recording and material culture inventory.

2.2 Material objects emerged from *gula apong* production activity

Next, the discussion will focus on the material culture that emerged from the *gula apong* production activity. The process begins with the preparation of the inflorescence stalk of *nipah* palm. The stalk can be chosen either from the young (*mayang*- flowering) or old (when the fruit is already beginning to form). But according to Yahya, the old inflorescence stalk is better since the sap can be harvested every day for three to four months. In the past, people shacked the stalk by hand and this process required tremendous effort and energy.

Figure 3 *Nipah* Palm Sap Tapping Process

Now, they beat the stalk using *pemaluk matuk* (a wooden beater) for the same purpose, to promote quicker and plenty of sap production. The process is repeated after three days for a month before the sap can be tapped and collected. The tapping process (*menyabit*) uses a specially made knife (*pisok penyabit* in Figure 3 inlet) to trim the stalk to remove surface crust and open the pores, and bamboo container (*buluh tuak*) is will then collect the sap. The tapping is done in the late afternoon and early in the next morning, the filled *buluh tuak* will be collected.

The local use *buluh minyak* or scientifically known as *Bambusa vulgaris* due to its properties, strong, light, straight stem, thin-walled and long clump interval. *Air tuak* spoils quickly (sour) due to the presence of water and impurities. Thus, the *buluh tuak* must always dry and clean. After use, the bamboo will be cleaned with *ngujak* (a brush made of *nipah* stalk) and dry and smoke on top of the hot clay stove.

Figure 4 Material objects used in *gula apong* production a) *buluh tuak*, b) *pemaluk*, c) *ngujak* and d) *ludan* and *solen*

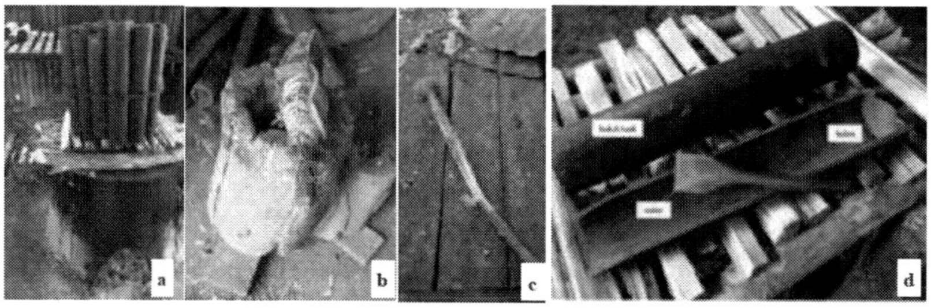

Figure 5 The use of (a) *solen*, (b) *gayung* and (c) *ludan* in *gula apong* production

The harvested *air tuak* is then brought to a *dango* (a processing plant) on the *nipah* palm site. The liquid pours into a big wok (*kah*) on top of a clay stove (*kui*) fire, using wood. A stirrer, made of coconut shell (*gayung*) to stir. *Enggalang* is used to avoid overflow of boiling syrup during the cooking process. In the past, *enggalang* was made of wood bark (preferably *Meranti* wood) which is now replaced by a custom made aluminium drum. *Gula apong* processing involves boiling the *air tuak* for several hours until the liquid is reduced by about 90% to a dark concentrated syrup that is sweet. The colour (yellowish, brown and dark brown) of the syrup is depending on the cooking time taken. The dark brown sugar taste bitter and inappropriate for cooking, but used to produce alcohol. When cooked, the *gula apong* is poured using *ludan* and *solen* (bamboo funnels and wooden spatula) into a container.

Figure 6 Cooking Materials

2.3Resources sustainability strategy

As a *gula apong* producer, a person is commonly owned up to 500 *nipah* palm trees. To ensure the sustainability of resources, the *nipah* field is divided into blocks. Thus, the tapping process will cover one block at a time to provide sufficient time for the trees from the other blocks to recuperate before they can be tapped again. The *dango* is built near the *nipah* field and water source so the everyday routine of *gula apong* production is managed efficiently.

Figure 7 (a) *Nipah* field, (b) water source and (c) *dango*

3 The continuity of traditional subsistence activity and settlement pattern

The finding suggested that the Malay of Saribas is heavily reliant to a river (Batang Saribas) which becomes the major influence that determine their settlement pattern and subsistence activities which then shaped the material culture assemblage. The river continues to be the main highway to connect the community to the other parts of Sarawak and also as a source of marine resources determine the choice of linear settlement pattern along the river and fishing activities. The results also suggest that brackish water river terrestrial resources; *apong* and *mulong* accessibility and abundance availability that influence the community in continuing the traditional subsistence activities (*apong* and *mulong* based food production) related to those vegetation sources. Consequently, the assemblage of material culture that exist within the Saribas Malay community is the representation and manifestation of their choice of settlement pattern and subsistence activities.

4 Conclusion

The regional settlement pattern and environment of Saribas Malay in the context of Malay in Kg. Tambak, Pusa, Betong demonstrate the importance factor that reflects (1) choice of lifestyle and their subsistence activities as the community's strategy to adapt to settlement pattern and environment, (2) variants and designs of their material culture, use and choice of material and best practice in material culture production, and (3) natural resources sustainability strategy.
The abundance of *nipah* palm within their settlement and ecosystem play a significant role in the development of technology to effectively utilize the natural resources in food production technology; *gula, garam, cuka* and *mayang*, and it also directly influences the emerging of material culture related to that activity.
The authors wish to acknowledge the following entity and individuals that had been providing support for this study in various capacities.

1. UNIMAS Special Grant SSTG, Reference No. - F03/SpSTG/1571/2017– for providing financial and equipment support during the period of research.
2. Mr. Yahya Keram, Mr. Rosman Basuni and Ms. Rosnah Kadir for their assistance in the field.

References

1. Awang Azman Awang Pawi, *Wajah Melayu Sarawak: Rampaian Ilmu Budaya* (pp 187-210). Kota Samarahan: Universiti Malaysia Sarawak.
2. Johnson, D. V., World Wide Fund for Nature, & International Union for Conservation of Nature and Natural Resources. (1991). *Palms for human needs in Asia: Palm utilization and conservation in India, Indonesia, Malaysia and the Philippines*. Rotterdam: Published for the World Wide Fund for Nature and for the World Conservation Union by A.A. Balkema.
3. Mary Fatima Subet dan Salbia Hassan (2014). Rahsia Kebahasaan Pantun Melayu Sarawak. In Awang Azman Awang Pawi, *Wajah Melayu Sarawak: Rampaian ilmu budaya* (pp 187-210). Kota Samarahan: Universiti Malaysia Sarawak.
4. Sabihah Osman, "Keadaan Masyarakat Bumiputera Sarawak pada Abad Kelapan Belas dan Awal Abad Kesembilan Belas: Satu Tinjauan Sejarah" dlm. Malaysia in History **XV**: 2, 1972.
5. Sanib Said, Penyelidikan Terkini di Malaysia, & Institut Pengajian Asia Timur, Universiti Malaysia Sarawak. (2012). *Sejarah Awal Kepulauan Melayu: Lima buah negeri warisan Sarawak yang hilang*. Malaysia: The Ministry of Higher Education Malaysia.
6. Sanib Said, 2013. Melayu Sarawak: Sejarah yang Hilang. Kota Samarahan: Penerbit Universiti Malaysia Sarawak.
7. Tsuji K., Sebastian L.S., Ghazalli M.N.F., Ariffin Z., Nordin M.S., Khaidizar M.I., & Dulloo M.E. (2011). Biological and ethnobotanical characteristics of Nipa Palm (Nypa fructicans wurmb.): A review. *Sains Malaysiana*, **40**(12), 1407-1412.

Community restructuring and sociocultural interaction identity of *orang laut (kuala)* in the state of Johor

Maryam Syafiqha Mohd. Sayuti[1], *Awang* Azman Awang Pawi[2], *Jamilah* Omar[3]

Department of Socio-Cultural, Academy of Malay Studies, Universiti Malaya, Kuala Lumpur, Malaysia [1, 2]
Faculty of Art, Computing and Creative Industry, Universiti Pendidikan Sultan Idris, Tanjong Malim, Perak, Malaysia [3]

Abstract : This study will look at how this process occurs among the *Orang Laut (Kuala)* in Johor. This study was carried out to analyse the restructuring of the community and the identification of indigeneous ethnic (Orang Laut) socialcultural interactions in Rengit, Minyak Beku and Kota Masai Johor. In particular, the study will involve aspects of the sociocultural life of the Orang Kuala Rengit, Minyak Beku and Kota Masai Johor, which include languages, economic activities, customs, and practices that are still practiced and things that have been modified in accordance with the factors around them. Analysing the process of adaptation is done closely by relating the process of interactions within groups, local communities, government agencies and the adaptation patterns that occurred. Besides that, the sustainability and the preservation of ethnic cultures of *Orang Kuala* in the social system as a whole and their impact on the development of the structure of community are examined. Qualitative method is used assisted by other study materials obtained such as field work, observation, ethnographic which are reinforced with library research such as documents and other materials related. This paper will cover the restructuring process by explaining the concepts, books and previous research reviewed by the researcher.

Key words: Orang Kuala, restructuring, sociocultural identity

1 Introduction

The problem of community restructuring has begun the past. Among the Indigenous peoples of Malaya, restructuring problems are often linked to life activities, social systems and cultural activities from generation to generation. One of the factors leading to community restructuring is to eradicate the poverty that occurs in a society, (Just Faaland, Jack Parkinson & Rais Saniman, 2002). This scenario also involves the Orang Asli community that is part of the indigenous community in Malaya and subsequently due to the expansion and transformation of the political administration system into Malaysia.

According to Giddens (2011), the backwardness and retardation of a society's life stems from the internal aspects of the society. Therefore, the real cause of the external factors of the community is not to dictate the traditional lifestyle system that prevents them from achieving sustainable living. Hence, this decline is due to the development of international economic networks following the occurrence of colonization in a society.

The concept of identity has been studied, analyzed and discussed by some scholars. Identity is a characteristic or unique or characteristic and personality for each individual as well as for a group (Giddens, 2008). In addition, identity is a set of meanings and experiences that contain an explanation of who an individual is or a group based on self-esteem and in the outsourcing of others (Giddens, 2008, Castell, 2004). Indigenous people in this study refer to the

*Corresponding author: maryamsayuti@gmail.com

Orang Kuala who have been said to have once made the sea their home in the shores of the Straits of Malacca Strait and Riau Islands. They are associated with the development of the Malay Sultanate of Malacca and Johor and has begun to change the culture of life, beliefs, and economics and become part of other groups of people who then affect the overall structure of the *Orang Kuala* community. Also states in many parts of the Sulalatus Salatin, the role of *Orang Laut* in the Straits of Malacca pictures acts as the straits gatekeeper. Caster (2004) argues that it is very important in society to have the meaning and experience built through the aspects of identity such as name, language, religion, culture and so on. It works differently between self and others, thus giving meaning to the existence of self and others which leads to socio-cultural formation.

We know of no people without names, no languages or cultures in which some manner of distinction between self and other, we and they, are not made…self-knowledge – always a construction no matter how much it feels like a discovery – is never separable from claims to be known in specific ways by others.

Hence through this study, the process of identifying the changes that occur can either can lead to the extinction of this society or the development in a different form with a change of system that leads to a new dimension of *Orang Asli* society. They will only remain a record in history which in actual fact, the role and formation of the early culture of the *Orang Asli* society is very important in the history of the development, civilization and heritage of the entire Nusantara region.

1.1 The Origin and Concept of *Orang Asli*

Like all ethnic groups, the *Orang Asli* community is divided into three (3) ethnic groups formed from 19 tribes. The community has a very unique cultural identity, rich in various aspects when discussing each ethnic group. They are a strong community with customs and values inherited from the previous generation (Juli Edo, 2000).

Orang Seletar, Duano and *Kanaq* are the 3 sub-tribes of the *Orang Laut* that can be found in Malaysia (more sub-tribes of *Orang Laut* or *Orang Selat* can be found in Riau Islands, Indonesia). These three sub-tribes have been combined with the Jakun and Temuan tribes as the Proto Malays, one of the three groups of *Orang Asli* (Aborigines of Peninsular - besides the Malays) recognized by the Malaysian government. Just like the *Orang Laut*, the Jakun and Temuan are also the sub-ethnics of the Old Malay, but they live on land. They are known as *Orang Darat*.

Orang Seletar can be found around the coast of Johor Bahru district. The name *Seletar* was taken from a place called Seletar in Singapore. Originally they inhabited the Tebrau Strait and the Singapore Strait before slowly moving (being moved) to Johor. This migration reached its peak in the early years of Singapore's separation from Malaysia. Today, the *Orang Seletar* in Singapore is declared extinct either due to death or assimilated with the Malay-ethnic. In Malaysia, *Orang Seletar* still continue to live their traditional way of life despite staying permanently on the coast. *Orang Seletar* was slow in accepting Islam whereby some embraced Christianity and Buddhism. The Seletar language, just like the catastrophe confronting their settlements, is slowly dying. It is drowned by other groups of languages that are more dominant around it. Although originally only 20% of Seletar's vocabulary is similar to Malay language, the Seletar children are now more comfortable speaking Malay.

Orang Duano originated from Bengkalis Island (Sumatra) before spreading to many other islands around the Straits of Malacca, the west coast of Johor and the south east coast of Sumatra. In Johor, the Duano inhabit the confluence of rivers in Batu Pahat and Pontian, but some of them were transferred to permanent settlements in Kota Tinggi. The language of the *Orang Duano* is different from Malay language, but it can be said that this language will soon be extinct as the young generation prefer to speak Malay as taught in schools. I have seen the lives of the Duano people in Kampung Bumiputra Dalam in Rengit, Batu Pahat and their assimilation with the Malays is clearly visible. From the information that I obtained, the younger generation not only do not understand the Duano language, but they are also ashamed to be labeled *'Orang Asli'*. For the purpose of distinguishing them from the Malays, they are more comfortable to be called *Orang Melayu Kuala*. In Malaysia, the *Orang Duano* are Muslims.

Orang Kanaq are the smallest sub-tribe of the *Orang Laut* from Malaysia and Indonesia. The number that can be traced to this day is only 87 people, all living in Kampung Sungai Selangi, Kota Tinggi, Johor. The findings of the field work also concluded that none of the Kanaq tribes in the Riau Islands were found even in their own place of origin in Sekanak Island, a small island in northern Batam. Based on historical records (The Sosioeconomic History of The *Orang Kanaq* of Johor, Mahani Musa, 2011), the *Orang Kanaq* were brought to the mainland of the Peninsular during the Sultanate of Johor for the purpose of supplying forest products. However, they failed to do so because they were not accustomed to living on land until they were neglected and unable to develop their own economy. On the welfare of the Johor state government, each family was awarded 10 hectares of land to work on besides one special village for them in Kota Tinggi (Kampung Sungai Selangi, now). However, the sea instincts in their lives were not easy to erode, the land that they were blessed were not cultivated and were leased to Chinese entrepreneurs. Living in a special village had also prevented the *Orang Kanaq* from exposure to the outside communities surrounding them. Since the time of the Sultanate of Johor, the *Orang Kanaq* had embraced Islam. Nevertheless, the traditional beliefs of their tribe still remain to this day. It also relates to the limited socializing of this tribe with the outside communities. The children of the *Orang Kanaq* are still reluctant to go to school, despite being given various assistance. According to a researcher, the *'Orang Asli'* label is increasingly isolating them from continuing to compete with the outside world. But there is one good thing, the Kanaq language is preserved to this day. The language is able to withstand the test of time and is spoken by all generations although its vocabulary is slowly being replaced by the Malay language vocabulary through the mass media.

Orang Asli is one of the indigenous tribes grouped in the native Malay tribe. According to Juli Edo (2000), the grouping was carried out by the British when administering Malaya as one of the efforts to overcome poverty problem in order to develop indigenous people. However, the term *Orang Asli* is often used and widely used by the Orang Asli Development Department (JAKOA), an agency responsible for *Orang Asli* affairs in Peninsular Malaysia. According to Mr. Osman Bungsu (2017), the term *Orang Asli* changes according to place, whereby in Johor, they are known as *Orang Laut* or *Orang Kuala* based on the location of their settlements whether at the sea shore or coast of Johor. It is said that this tribe originated from the coastal areas of Riau, and this is different from the *Orang Asli* community in other states such as in Pahang and Kelantan, where the *Orang Asli* are much more populated in the interior of the districts in these states.

Orang Kuala was a title given to them in Johor. Originally, this group was known as *Orang Laut*, a group of 'indigenous tribe' in Riau. *Orang Kuala* or in their tribal tribes known as the Duano tribe are one of the 18 tribes of the *Orang Asli* in Malaysia. They belong to the Proto Malay group, one of the main groups of the *Orang Asli*. In the Malay historiography, it is also stated that the Proto Malay communities were living in the southern region of the peninsula of Malaysia and we cannot dismiss that this group was in Johor.

Tracing the history of Duano tribes in Rengit, Minyak Beku and Kota Masai is not easy because there are not many written documents explaining the arrival of these tribes in Malaya or Malaysia. According to Nazarudin Zain, 2015, based on interviews with the old generation, there are various opinions regarding the history of the Duano tribe in Benut and the history of the origins of this group. Some have argued that the settlement of the Duano tribe in Benut has existed since 1928 and their initial settlement was located at the Benut River. The title given to them, *Orang Kuala* is due to the location of their settlements near the confluence of rivers or the coastal area.

They are widely placed in the Johor region like Pontian, Rengit and Minyak Beku. This is because of the location of settlements which are at the confluence of rivers and coastline near the rivers. Once upon a time, this group was evenly distributed on the beaches on both sides of the Straits of Malacca, including the Malay Archipelago.

In the context of our country, although they are called *Orang Laut* but the Malaysian government categorized them as *Orang Asli*. This is enshrined in Act 134 which gives the privilege to the *Orang Asli* community and on the administrative aspect facilitates the development of this community.

The tribesmen of *Orang Kuala* do not like being called *Orang Kuala*. In their view, the term is usually used by outside communities to mock them. Hence, they are more comfortable to be called *Orang Laut* or *Orang Asli* or *Melayu Asli* and some favour to be known as *Orang Duanu*. But in fact, the tribemen of *Orang Kuala* are a sub-race to the Malay Proto / Asli. They are known as *Orang Kuala* due to their earlier settlements which were in the confluence of rivers and lives as fishermen. The tribesmen of *Orang Kuala* often call themselves *Orang Laut*, while the *Orang Asli* on land as *Orang Asli Darat* or *Orang Sakai*.

The *Orang Kuala* are indeed known as *Orang Laut*. However, in the large group of *Orang Asli*, besides *Orang Kuala*. *Orang Laut* also refer to *Orang Seletar* and *Mah Meri* (or also *Ma 'Betisek* or *Besese*). All *Orang Kuala* are Muslims, interestingly, they are different from other *Orang Asli*, the *Orang Kuala* are not new to Islam, but they are Muslims since their ancestors' time. In fact some of them regard themselves as the Malay race on the grounds that they are not converts, but the original Muslims as the Malays. Some older people also claim that they have a bond with the indigenous people in Riau Indonesia. However, if based on the views of Chou (2003) and Andaya (2008), the indigenous people in Riau are not regarded as *Orang Kuala*, but are called *Orang Laut* or *Suku Sampan*.

1.2 *Orang Asli* in the History of the Malay Civilization

The role of the small Malay tribe such as *Bajau* and *Orang Laut* is actually big in history. In the writing of Mohd. Affendy Abdul Hamid in *Berita Harian* dated 13 October 2010, he explains that after examining the story of *Orang Bajau*, the nomad tribe and sea traders in the waters of the Philippines, Malaysia and Indonesia, *Orang Laut* were mentioned in the Malay History. These tribal people served the Sultan or the Malay rulers by protecting the sea route.

In 1819, according to British records, Singapore was described as a fishing village. In other words, before 1819 Singapore was not so important. Yet this is not the case. Archaeological studies (excavations) by the historian of Singapore National University, Dr. John Miksic, have exposed the important role of Singapore during the 14th century. This means that Singapore's existence as a trading port has long been in existence since the ancient times. In fact, Thomas Stamford Raffles realized this when visiting Singapore in 1819 to re-establish it as an entrepot port. Dr. Leonard Andaya, in Chapter 6 of the book entitled *Leaves of the Same Tree: Trade and Ethnicity in the Straits of Melaka* (University of Hawaii Press, 2008), unveils the role of the *Orang Laut* which was very important in the formation of the Malay kingdoms (spelt as Malayu and not Malays). *Orang Laut* were regarded as important to be given the honorary ranks by the Malay rulers. They controlled the sea route and at the same time ensured peacefulness which was important for international trade.The *Orang Laut* were described as very knowledgeable about the sea in terms of seasons, waves, seafood such as fish and the pearls needed for the Chinese market. However, their biggest contribution to the Malay rulers was to preserve the sea route including 'persuading' the merchant ships to stop at the port. Nevertheless, the *Orang Laut* devotion decreased after major changes occurred in the development of the Malay economy from relying on the sea to agriculture and mining beginning from the late 19th Century AD. In fact, there are very few research on the *Orang Laut* in terms of their history, culture and thinking. Their presence was to guard the coastal kingdoms of the Straits of Malacca, including the Sultanate of Malacca. Based on their 'isolated' way of life and occasional encounters with people, we need to look into their perceptions about, for example, developing the Malay language and culture in Malaysia and Singapore; the overall impact on the structure of the *Orang Laut* community since Singapore's separation from Malaysia; the role of Chinese in the history of Southeast Asia - how did the *Orang Laut* perceive the Chinese community?; How do the *Orang Laut* adapt to the current Southeast Asian transformation and what are the *Orang Laut* perspectives about these processes? In an article from *The Straits Times* dated December 25, 1949, on the Singapore's *Orang Laut* in the south of Johor, the Chief of the *Orang Laut* was called '*batin*' or '*tok batin*'. At that time, there were two tribes of the *Orang Laut* in Singapore, known as *Orang Kalang* and *Orang Selat*. *Orang Kalang* came from Riau and settled on Kalang Island, but they had to move when the airport was built at that time. Many *Orang Kalang* live in Pasir Panjang. And the *Orang Kalang* group is divided into two, namely *Orang Kopek* and *Orang Senggera*.

The *Orang Selat* are said to be followers of Temenggong Abdul Rahman and live on their boats at the estuary of the Singapore River. They acknowledge themselves as the tribe sailing the Raffles' boat to land on the banks of the Singapore River. The description of the economic activities of the *Orang Kalang* is indeed interesting. They sell young nipah leaves to make cigarette papers. They collect the nipah in the Riau Islands using small sailboats called '*perahu pucuk*'.

M. Effendy A. Hamid goes on to say, though the *Orang Laut* are said to be uneducated or are said to shy away from progress, these are not true. Some are able to adapt to change, persistent and profoundly perserve their dignity, and proud of their origin as 'Singapore's *Orang Asli*' like the former cabinet minister, Haji Othman Wok. Being addressed as *Orang Asli* is not suitable with their community structure. This makes them unhappy to be categorized as *Orang Asli*. According to them, there are several factors why they refuse to be categorized as *Orang Asli*. First, they are Muslims but the *Orang Asli* are not. *Orang Asli* is also known as *Orang Asli Darat* and most of them practice Christianity. In fact, they claim to be Muslims and should be considered as native to the local Malays. Secondly, with regard to the Islamic religion they are prohibited from eating pigs and rearing dogs whereas the *Orang Asli* community can eat pigs and are allowed to rear dogs.

It is wise to study this before the story of the *Orang Laut* is no longer known and this tribe becomes extinct.

2 Research Methodology

This study was conducted using qualitative approach, which emphasizes the quest for meaning and process. Stephen & Kasim (2015) explains that qualitative research is the process of digging the understanding of meaning and describing the reality investigated in depth and is multifaceted, complex and involves development of theories. The use of qualitative approach is based on the determination to understand and do an indepth review of the process of structural change in society and the association with the determination of the identity of the socio-cultural interaction of society which on the verge of climishing. The collected data will prove to be the precise truth, strengths and weaknesses to the explanation provided.

The study was carried out in several *Orang Asli* villages of the *Orang Kuala* or Duano in Batu Pahat, Johore and specifically at three main locations namely Rengit, Minyak Beku and Kota Masai. The selection of these locations was made because of the uniqueness shown which are different from other indigenous people. Among the uniqueness shown are the family structure and their socio-cultural concepts of transforming their economy, education and life through participation in the economic activities of the people of Johore. *Orang Kuala* Rengit and Minyak Beku have participated in the import of used goods while the *Orang Kuala* in Kota Masai usually carry out sea-based economic activities which are selling of marine products in the morning and evening to the local community. Indirectly, they have opened many job opportunities to their family members as well as to the local communities.

Through the use of qualitative approaches, researchers also obtain research materials using various means, which involve references to written abstracts (historical books, archives). Rich data collection in the field at the three main locations specifically in Rengit, Kampung Minyak Beku and Kota Masai in Johor are information on the similarities dan differences among them although at a glance they may look alike. The collection and analysis of written materials are very important to obtain a comprehensive picture, revealing the historical beginning of the existence and development of this society in Malaya and the Archipelago. Selection of locations through field-based research is focused on the role of individuals in the community, the position, the layers of different generations obtained through interviews and observations.

Interviews were conducted with every *Tok Batin* of the *Orang Kuala* at each place, several family members, resident associations and a few traders involved in their economic activities. Both structured and independent interview methods are used repeatedly by providing questions related to the role and position of informants in their groups. Researchers also took repeated observations to ensure that the validity and reliability of the phenomenon seen is a real phenomenon and not momentary. This is because the uniqueness shown is different to the *Orang Asli* community as compared to other societies. Repetition of interviews and observations also help researchers to organize and conduct effective trangulation checks during data analysis process and discusion. This is to assist the researcher to familiarize with these community and to promote mutual friendship to ease the sharing of information.

3 Discussion

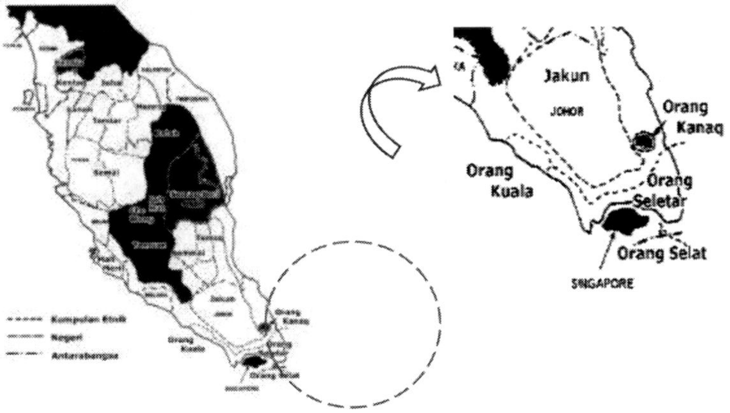

Figure 1: The Settlements of *Orang Laut/Kuala* in Johor

3.1 The Development of Sociocultural Aspects of Life

The discussion focuses on the 3 villages studied. The map in Figure 1 shows that the location of distribution of the *Orang Kuala* is very closely related to the sea and indirectly their lives cannot be separated from the sea and sea products. In the past, the *Orang Laut* community lived in a type of boat house called *'sampan panjang'*. However, they often move from one place to another to preserve sustainability of their lives which are self-reliant in nature. At the beginning of the *Orang Kuala* historiography, they were placed everywhere in the form of small settlements based on the tribe or group of their family members. However, after Malaysia achieved its independence they were then placed in well-organized and structured village areas which were provided by the government under the responsibility and administration of the Department of Orang Asli Affairs (Pelras, 2002).

Analyzing the structure of society and the Sociocultural Interaction Identity of this community can be viewed through three main aspects namely the development of languages and dialects, economic activities, customs, religions and practices, and changes in locality of settlements and interaction patterns. Overall, the growth and development of an area influence the changes in the structure of the community and consequently affect the sustainability of sociocultural identity. If viewed through the timeline, changes in the sociocultural aspects are quite different in comparison with the original structure of the community.

3.2 Development of Language and Dialect

In general, language is a system of sound symbols used by members of a community to collaborate, interact, and identify themselves. The language function can be divided into two, namely the general language function and the specific language function. The general language function includes as a communication tool, as a means of expression and as a means to adapt and integrate socially. While the specific language function is aimed at the practice of a culture, and most importantly is to gain knowledge.

Dialects can also be associated with variations of a group of speakers located at a place, region, or area. In other words, dialects are uniformed ways of pronunciation or styles of language use. The variances in dialects occur due to social and regional differences.

As such, the aspect of language and dialect is also one of the important aspects to be observed in a community. However, those called *Orang Kuala* usually use the same Malay language and dialect as the local Malay community in Johore. They are more comfortable using Malay language than the dialect of language they have practiced since long ago. They only use their language and dialect when they talk to each other to facilitate understanding and instruction they want to achieve.

3.2.1 Orang Kuala Rengit

This community group lives on the edge of Sungai Rengit, in the confluence of the Batu Pahat River. Their village area is formed in line with the direction of the river. The Village Head is appointed by the local authorities with reference to 'their elders'.

Most *Orang Asli Rengit* interact with the Malay and Chinese communities around Batu Pahat Town. Interaction with the local communities takes place due to the aspect of education, where they send children to schools around the area. Many Chinese traders around Rengit area influence the development of this community. The use of Malay language and Chinese dialect in everyday spoken exchanges has altered their language structure and reduced the language use.

This change is also due to intermarriages between the *Orang Asli* community and the Malay population in that area. Young people who are mostly exposed to external culture and social media have influenced the retention of the language of this community. The use of telephones, television station networks, and the internet brings a lot of impacts to the younger generation in particular.

3.2.2 *Orang Kuala Minyak Beku*

The use of their language has started to decrease due to their increasingly scattered settlements, usage of Malay language in daily interactions either in schools, communities or with other community groups. The Minyak Beku community are not much different from the people in Rengit, because of the geographical structure that places these two settlements in parallel. If in Rengit there is settlement separation, in Minyak Beku almost 70% have lived in Malay villages. Intermarriages between the Malay community and the *Orang Kuala* are no longer strange.

3.2.3 *Orang Kuala Masai*

Based on observation, the *Orang Kuala Masai* are still using their language and dialect in their daily activities. On the other hand, the *Orang Kuala* in Rengit and Minyak Beku have long been using the local Malay language because most of their residences are located near each other, and it is also easier to interact with the local communities.

4 Economic Activities

During the 70s, their main economic activities were fishing, and becoming fish retailers. However, with the development of the Rengit town, the economic activities have changed. They began to change their economic activities, the Johore Government also encouraged them to run the bundle business of household necessities such as furniture, electrical goods, kitchen utensils, clothing and toys. Initially, they started businesses on a small scale in their settlements and were just doing business in front of their houses. The *Orang Laut* have the unique identity of a community who cherish, adore and are close to the sea, symbolizing the 'who and the origins' of the sea as a source of livelihood and economic resources to them. The seas surrounding the Straits of Malacca have rich resources and become part of the development area of the community system. However, after the Department of Orang Asli Affairs has taken on the role of managing this community, they have been given the privilege by the Johore State Government to help stimulate the development of the *Orang Asli* community by promoting economic activities in a particular area.

4.1 Orang Kuala Rengit

The *Orang Kuala Rengit* showed an increase in their economic activities in line with the community's development. Compared to the 1980s, the community have now advanced as traders and some still remain as seafarers. The demand for fish from the local market makes this inherited economic activity a priority to them.

In fact, some of them are active in running workshops of used goods in Kampung Bumiputra Dalam, Rengit Batu Pahat (in Figure 2). They have long been involved in economic activities and at the same time help to increase their family income. Most of them run businesses such as selling imported furniture from Singapore which are in good condition and can still be used (in Figure 3).

Figure 2: The Entrance to the Village of the *Orang Kuala Rengit*

Figure 3: Used Furniture Sales Managed by *Orang Kuala Rengit*

4.2 Orang Kuala Minyak Beku

A visit to the Minyak Beku area showed that they have moved their settlements from the Batu Pahat riverbank area to the village area. The opening of quarry and shipbuilding companies as well as small ports in Minyak Beku has changed the structure of their settlements. However, most who were seafarers have been working with the Chinese and some are still the buyers or sellers of fish in the market. Several tables in the market are run by the persistent *Orang Kuala* sellers along with the Chinese and Malay sellers. What is interesting, if the mother is selling, then the children will also take part and they will not attend school. Some are also involved in the business located in the *Orang Kuala* Market at Jalan Pasar (behind the public market in Figure 4). They place their bundle sales daily from 8am until 12pm at Jalan Pasar. The bundle sales run by them include selling second-hand electrical goods at reasonable prices (in Figure 5).

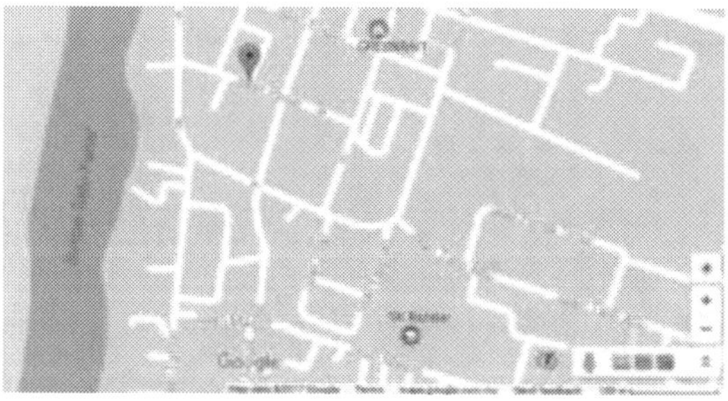

Figure 4: Jalan Pasar

Figure 5: Bundle Sales

They are no longer involved with economic activities in Batu Pahat. This is because they accept employment as government employees as their permanent jobs. They are more comfortable when grouped together with the local Malay community.

4.3 Orang Kuala Kota Masai

The Kuala Kota Masai village is located at the end of the Johore causeway in Pasir Gudang, near the Johore Fisheries Department's Fish Collection Center, the Y-shaped junction separates the Malay village and the *Orang Seletar* village, in contrast to the people of Kuala Minyak Beku that live among the Malay community. The main focus of this community is to become seafarers. In the village, 76-year-old Tok Batin Nah bin Bechik is the second *tok batin* after his father. The village is located on the edge of the Tebrau Strait. The economic activity of this community is selling of fish hauled.

According to William, the son of an *Orang Selatar:*

"Setiap pagi kami akan ke laut untuk menangkap ikan. Kebanyakannya kami akan dapat ikan yang besar-besar seperti ikan merah, ikan siakap, ikan pari dan banyak lagi. Setelah pulang, kami akan agihkan ikan tersebut kepada peniaga makanan dan selebihnya kami akan membuat jualan di pasar-pasar besar dan di pasar Orang Kuala Kampung Masai."

("Every morning we will go to the sea to catch fish. Mostly we will be able to get big fish like snapper, seabass, stingray and many more. After returning home, we will distribute the fish to the food traders and the rest will be sold in the big markets and in the *Orang Kuala* Kampung Masai market.")

P1,KKM,2

This is proven by the sales of fish operated by themselves in some areas such as their own village and some neraby markets (in Figure 6).

Figure 6: William (left - in red) along with family members running their business of selling sea products every 5 pm at the entrance of their village in Kota Masai.

"Ikan tak macam dulu, dulu banyak sekarang dah berkurangan, pembangunan di tambak menjadikan bekalan sumber laut berkurang…macam saya sebagai Tok Batin dapat elaun manakala yang lain masih menjadi nelayan."

("The amount of fish has now decreased, the development of the causeway has reduced the supply of marine resources ... for me I get allowance as a Tok Batin unlike others who remain as fishermen.")

P2,KKM,2

The government also provides a site to enable them to sell their catch. The sales area is on the outskirts of their village. Based on observation, the price of fish sold is cheaper than in the market. Among the catch they often get is grouper, red snapper, seabass, big eyes, snapper, cockles, crabs and promfret. Their catch is also reserved by restaurants around Johor Bahru. Some of them have their own vehicles like William's father who has become a supplier to several seafood restaurants. The demand among buyers is still encouraging so they are still fishing in the sea and selling their catch at a good price.

b. Customs, religions and practices

According to informants, the migration of the *Orang Laut* from their ancestors in Riau and islands around the Archipelago is similar to the Malays in terms of culture, religion, customs and language. The language of this group is known as *Orang Laut* language which includes various types of tribal accents. They generally seek livelihood by working as fishermen. Apart from being characterized as a group linked to the sea, they are identified as Muslims.

They are also known as *Orang Lanun* or *Orang Selat* because before they devoted their loyalty to the Srivijaya Kingdom, they were originally a group of sailors (some authors named them pirates) who guarded the area around the Straits of Malacca. They controlled the boundaries of the sea, chased out pirates, led the merchant ships to the destination port, took care of business activities and became the messenger for governments around the Straits of Malacca. Due to the position of *Orang Laut* who knew the intricacies of the sea around the Straits of Malacca described in the History of the Malays to be among the busiest sea route, they played an important role in the reign of the Kingdom of Srivijaya, the Sultanate of Malacca and later to the Sultanate of Johore-Riau. Their duty as warriors or police in the sea led to a close relationship between the *Orang Laut* and the three royal institutions that had been established since the early days of the reign of Srivijaya kingdom. *Orang Laut* were also known as a very loyal group to the royal family of Srivijaya. Even after the fall of Srivijaya in 1025, *Orang Laut* continued to be devoted to Parameswara, the last prince of Srivijaya (Palembang). Changes in the Johore-Riau sultanate system in 1699 also occurred when the 11th Sultan Mahmud Syah died as a sultan who was still single, then the throne was replaced by the Bendahara who appointed himself as the 4th Sultan Abdul Jalil which caused the *Orang Laut* not to recognize the new Sultan of the Johore-Riau Sultanate. Finally, by the 18th century, the role of the *Orang Laut* as the peace keepers of the Straits of Malacca was slowly being replaced by the *Orang Bugis*.

4.4 Orang Kuala Rengit

Their customs, religions and practices have changed considerably in particular due to changes in religious beliefs. Some people of the *Orang Kuala Rengit* community are also Muslims, in the village of Kuala Rengit a surau was built since the 1990s. The Malay custom became their custom as they celebrate all the festivals of the Malay community.

4.5 Orang Kuala Minyak Beku

The Minyak Beku community are not much different from those in Rengit, because of the geographical structure that places these two settlements in parallel. If in Rengit there is a settlement separation, in Minyak Beku almost 70% have been living in the Malay village. Intermarriages between the Malay community and the *Orang Kuala* was no longer strange. There are already those who wear Islamic clothes with veils, covering *aurat* and practicing the cultures of the Malay.

4.6 Orang Kuala Kota Masai

The settlement of this community differs from the previous two settlements, their location is quite isolated and far from development. Judging from the customary aspect, they are still bound by the family structure and the *'tok batin'* as the leader. Tok Batin Nah bin Bechik has determined his successor. In this community, Tok Batin Orang Kuala Kota Masai has nine (9) children and two (2) of them have married the Chinese and their children use the Chinese name in their identity card. In the presence of the researchers, they were actively discussing the 'modern *sewang*' event called the joget night which will be held in Setulang, the village of the *Orang Kuala Selatar*, in Johore. Although some of these families are Muslims, the Islamic characteristics are very few and the emphasis on Islam can only be seen on identity cards. Some even have adopted other religions as a result of mixed marriages with the local Chinese.

5 Changes in the Location of Settlements and Interaction Patterns

5.1 Orang Kuala Rengit

For the *Orang Kuala Rengit,* the development provided by the government has given them a lot of benefits. Former Senator Mr. Othman Busu in an interview said the government always provided an opportunities for the *Orang Kuala* community to carry on business especially around Bandar Rengit and Batu Pahat. Mr. Othman himself is a supplier of teak furniture to the *Orang Kuala* traders. They are also traders at the *Pasar Orang Kuala* in Bandar Batu Pahat by offering clothes, home appliances, furniture, and reusable decorative items. The demand for goods turned them from fishermen into traders. Nonetheless there are still those who remain as fishermen by providing fresh fish to the locals. The impact of this economic activity led to a wider interaction patterns, with the Malay language as the medium of their communication.

5.2 Orang Kuala Minyak Beku

The community in the outskirts of Batu Pahat are affected by the rapid development of the city. Interactions with the Malay, Chinese, Jawi Peranakan (Arabs) and Indian communities are very clear due to the location of their settlements and also economic activities. In addition, their involvement in the public market of Batu Pahat as part of the traders shows a more complex pattern of interaction with other communities.

5.3 Orang Kuala Kota Masai

The Kuala Seletar community hardly mix with the outside communities, but they deal with the same buyers when selling fish, purchasing household supplies, which involve them interacting with outsiders. Their relatives live in other villages because according to Tok Batin Nah bin Bechik, each village can only have a leader, so every village is usually inhabited by a nuclear family.

6 Conclusion

Studying the restructuring of the *Orang Kuala* community by analyzing it specifically into interaction patterns, economic activities, and religious practices shows a minority group of community who are beginning to become extinct.

From the aspect of the structure of community, the change of settlements, development and practice of intermarriages have begun to break through the walls of this community which has weaken the traditional values, customs, and language. The birth rate in this group of community is also shrinking, if the head of the family or *Tok Batin* has nine (9) children, he states that his father has 19 children, but his own children have only three (3) or two (2) children. Similarly, the *Orang Kuala* at Rengit and Minyak Beku, almost all showed a decreased family size.

If viewed from the beginning it is imperative that as a minority community, they need to foster interactions with the local communities to secure their position in the world-view. From the perspective of the use of parent language, almost all new generations are not fluent in their mother tongue. Most have been using the Malay language as formal education is received from schools and through interactions with the local communities.

In terms of economic activities, however, after a long time as seafarers, having mastered and understood the sea but with the changes of development and technology, and lack of marine resources, have made them begin to move on to land activities. Furthermore, with the supervision of the Johore state government, many business opportunities are provided for them to show their potential in the economic field as they have been allowed to trade in an area reserved for them especially in Batu Pahat. That area is located in the Batu Pahat market and their businesses are still ongoing until today. Obviously, their potential in the economic field is extensive and good.

The continuity and sustainability of the *Orang Kuala's* culture as we see in the communities of other countries such as the Aborigines of Australia, and the Red Indians in the United States have had a profound impact from the development and technology of the modern world.

References

1. Andaya, L. A. *Leaves Of The Same Tree-Trade And Ethnicity In The Straits If Malaka* (2008).
2. Castell, M. *The power of identiti*. Edisi ke-2 (2004).
3. Chou, C. *Indonesian Sea Nomads Money, Magic And Fear Of The Orang Suku Laut* (2003)
4. Giddens, A. *Sociology* (2008).
5. Giddens, Anthony. *Teori Strukturasi : Dasar-dasar Pembentukan Struktur Sosial Masyarakat* (2011).
6. Google Maps. Maps of Jalan Pasar retrived on 14 April 2017 at https://www.google.com/maps/place/Jalan+Pasar,+83000+Batu+Pahat,+Johor,+Malaysia/data=!4m2!3m1!1s0x31d056c6a2e06537:0x228c075936af601f?sa=X&ved=0ahUKEwiQkpzRiKPTAhWDwI8KHcHsB9wQ8gEIITAA dan maintained at 2017 (2017).
7. Juli Edo, Dari darat ke laut: Singkapan ringkas etnosejarah Orang Asli di Johor.At Seminar sejarah dan budaya Johor, anjuran Kerajaan Negeri Johor, 3-6 Mei 2000. (2000).
8. *Just Faaland, Jack Parkinson & Rais Saniman, Dasar Ekonomi Baru: Pertubuhan Negara dan Pencapaian Ekonomi Orang Melayu*. . (2005).
9. Pelras, Christian, (*A few jottings on some aquatic populations of the Nusantara Archipelago*. Retrived on 14 April 2017 at Http://www.keene.edu/library/OrangAsli/ pelras.pdf. (2002).
10. Mahani Musa. *The Sosioeconomic History of The Orang Kanaq of Johor*. Kajian Malaysia, **Vol. 29**, No. 1, 2011, 47–74. (2011).
11. Siti Rahmah Wahab. *Adaptasi Sosiobudaya Orang Laut di Kecamatan Tanah Merah, Kabupaten Indragiri Hilir, Riau, Indonesia*. Thesis published. (2015).
12. Stephen Kwadwo Antwi & Kasim Hamza (2015). *Qualitative and Quantitative Research Paradigms in Business Research* (Journal of Business and Management [Online]) **Vol.7**, No.3, 2015 217. Retrived on 14 April 2017 at www.iiste.org. Tamale: School of Business, Tamale Polytechnic.
13. Zainun, Nazarudin. *Antropologi Dan Sejarah Dalam Kearifan Tempatan*. (2015)

Panglima Laot: His Legacy and Role in Conserving Marine Resources in Aceh, Indonesia

Muhammad Adli Abdullah[1] · *Azmi* Arifin[2] and *Sulaiman* Tripa[3]

[1]Syiah Kuala University, Adat Department, 23111 Banda Aceh, Indonesia
[2]Universiti Sains Malaysia, History Section School of Humanities, 11800 Penang, Malaysia
[2]Syiah Kuala University, Law and Society Department, 23111 Banda Aceh, Indonesia

Abstract. This paper discusses the significance of a historical heritage and its practices based on local wisdom to develop sustainable marine resources in Aceh. Local wisdom can be translated through the examination of the significant role played by the *Panglima Laot* (Sea Commander) as the enforcer of the Customary Law of the sea in Aceh. This law exists since the glorious era of the Kingdom of Aceh from the 17th century. The role played by the *Panglima Laot* underlines local wisdom in managing and ensuring the sustainability of the seas, an important economic source for the coastal community in Aceh. The sea is the lifeblood of the Acehnese coastal communities. From a historical point of view, *Panglima Laot* played a major role in safeguarding and nurturing the sea for the benefits of marine activities and maintaining the marine environment and its natural state, as a common property that should not be destroyed for the good of the future. *Panglima Laot* also managed marine resources that can be exploited and empowers the population to undertake marine activities in accordance with the customary requirements and the interests of society as a whole. In ensuring the effectiveness of the functions, the role of Panglima Laot is strengthened by the existence of special rules concerning marine and fishing practices established for generations through *Adat Laot* (Aceh Customary of the sea) to ensure the long-term harmony and sustainability of Aceh's marine resources.

1 Introduction

This paper aims to explore the legacy and the role of *Panglima Laot* in preserving the marine resources in Aceh. This paper is significant because *Panglima Laot* is a traditional Aceh fishery institution that is a legacy of the Aceh civilisation in the past, that but remains alive and continue to evolve to this day. *Panglima Laot*'s role is to enforce the Customary Law of the sea in Aceh which signifies the wealth of local wisdom; it remains contextual for the sustained development of fish resources in Aceh.

Beginning with this interest in mind, the discussion of the existence of *Panglima Laot* in Aceh cannot be separated from a number of aspects, namely history, anthropology, and law. History will describe the journey of the *Panglima Laot's* existence from the 17th century or earlier through to the colonial period and up to the present day. Anthropologically, it will illustrate how the progress of every scene in the journey involving *Panglima Laot* and how they play a role in sustainable marine development. From a legal aspect, it will trace how the Indonesian Kingdom recognises the existence of the Aceh Customary Law of the sea which is based on local wisdom, operating within the Indonesian national legal system which was based on the European Continental system.

The embodiment of *Panglima Laot* must be seen from a number of aspects since all these aspects define its existence to this date. Up till now, the empirical manifestations can be traced from the dynamics of its development from time to time [12]. Over a span of time, the development can be found to either strengthened or weakened. It is similar to the condition of the country in tandem with the developments in politics and law [15].

There are at least two important dynamics that occurred in the current Aceh socio-politics regarding *Panglima Laot*:

First, by placing *Panglima Laot* as a legacy of the traditional institutions of the Kingdom of Aceh heritage who solely functions in the celebration of *adat* (customary law) coastal communities. In this context, besides *Panglima Laot*, there are also a number of other customary institutions at the *desa* (village) and *mukim* (village federated) such as *Keujruen Blang* (Institution of agriculture Custom), *Pawang Uteun* (Institution of Forestry Custom), *Haria Peukan* (Institution of traditional market Custom). *Imum Mukim* (head of village federated). The embodiment of customary institutions is only focused on the concept of maintaining human balance in the community life in Aceh.

Second, ecological implications, where customary institutions become an important part of maintaining the ecology of the coastal and marine spaces in Aceh. In the coastal communities of Aceh, each marine customary territory has its own customary ecological space, which until now has not received full legal recognition.

The embodiment of *Panglima Laot* is related to the preservation of marine resources in Aceh, so it is impossible to ignore this ecological aspect. The necessity of maintaining ecological aspect in anthropologically explained through the concept of *ulayat laot* (tenure right of the sea). In the study of anthropology, the discussion on the *ulayat laot* started in the 1970s [9, 20, 18]. The concept of marine ownership, Indonesia subject to Dutch legal system which bound by the *Mare Liberium's* teachings that the sea should not be owned by anyone [11]. *Mare Liberium's* teachings were developed in the 17th century in the Netherlands. While England, Spain and Portugal embraced the *Mare Clausum*, which states that the sea can be owned. Both of these teachings (*Mare Liberium* and *Mare Clausum*) are essentially the same as the theory of *res nullius (mare clausum)* and *res communis (mare liberium)*.

Although the state did not fully recognise the *ulayat Laot* rights of the sea, the Aceh coastal community continue to practice it as the legacy of the Kingdom of Aceh. In this case, Adhuri states that the practice of *ulayat Laot* rights or the management of traditional marine resources continues to be applied in Indonesia as an alternative maintenance practice [9].
The *adat* management needs to be brought closer to the community in order to maintain the *adat* ecological space. The existence of *Adat* ecological space determined by how the government's policy administered [14]. Furthermore, ecological space allows this type of local wisdom to be used in the context of marine resources management [19].

The government policy should put priority for coastal community to participate in coastal management based on the *adat system*. *Panglima Laot*, as an *adat* institution in Aceh which is related to the rights to manage the sea resources in Aceh has very important role to support the government program in managing its ecological space [17]. The success of managing this *adat* ecological space will determine the role of the *Panglima Laot* in preserving sustainable fisheries in Aceh.

2 Adat Laot And Panglima Laot

In terms of the *adat* ecological space, there are two very important aspects that needed to be studied. First, the regulatory aspect of the state law. In Aceh, following the peace talks between Indonesia and the Aceh Mardeka Movement in Helsinki on 15 August 2005 in Helsinki, a new law was established namely the Republic of Indonesia's Law 11/2006 on the Government of Aceh, hereinafter referred to as the *Undang Undang Pemerintah Aceh* (UUPA) which also acknowledges the existence of the local wisdom of *Panglima Laot* (article 98 and article 165 UUPA), in the utilization of natural resources, should not dominate one another, in this case between economy, social, ecology, and cultural interests. To impliment the local wisdom which is stated in new law of Government of Aceh, the Acehnese parliament has passed the *Qanun* (by-laws) relating to customs and customs development, namely *Qanun* Number 9 Year 2008 and *Qanun* Number 10 Year 2008. In addition, the provisions about local wisdom can actually be traced through *Qanun* Aceh Number 7 Year 2010 on fisheries and marine affairs.

Second, from the position of sociological legal arrangement, there seems to be an attraction of vested interest in the management of coastal resources. This vested interest of coastal resource management based on local wisdom had often contrary to the overlapping regulation of the State law. This condition gives rise to the question of whether the position of local wisdom in the State law is only meant to accommodate global consensus, which the involvement of the communities in saving the environment as has been discussed at the Earth Summit in Rio de Janeiro in 1992, resulting in the Rio Declaration, in Principle 22 which states that "... indigenous peoples and their communities, and other local communities have an important role in environmental stewardship and development because of their knowledge and traditional norms".

Globally, it is already understood that local wisdom cannot be ignored. The concept of local wisdom is synonymous with local knowledge, indigenous knowledge, traditional ecological knowledge, and rural knowledge. According to Gadgil, Berkes, and Folke, local knowledge is a cumulative knowledge and hereditary belief of the community in their relationship with the environment [10]. In Indonesia, a concept was developed by Keraf, who regarded local wisdom as forming patterns of everyday behaviour towards fellow human beings, nature, and the Unseen. Keraf calls the practice of managing environmental resources as positioning the environment within a sacred domain [3]. This view of the environment is the special characteristics of people living with local wisdom. beside, they have owned what has been termed as the rights of the *ulayat laot* [22]. In today's modern era, local wisdom valued like *Panglima Laot* can be implemented for development program policy [7].

In Aceh, especially after the tsunami, the adaptation process has started to be fiercely rekindled [8, 32]. Practically, the inclusion of the *Panglima Laot* institution as a guardian of local wisdom continues to be challenged, especially when it involves protecting marine resources from the use of destructive fishing gear [23,24].

As stated above, the *Panglima Laot's* existence is closely related to the rights of the *ulayat laot's*. The term 'the rights of the *ulayat laot*' may be defined as a special power under the customary law for the management of the marine territory. The use of the term '*laot*' (sea) in the phrase 'the rights of the *ulayat laot*', is to distinguish it from the "the right of the *ulayat darat* (land tenure)". In fact, there are differences between *ulayat darat* and *ulayat laot*. *Ulayat darat* is more easily determined compared to *ulayat laot*. The *ulayat laot* are communal in nature, to regulate, to control, and exercise for the common interest; although it does not mean closing the opportunities for outsiders to access to the sea. The rules relating to the management of rights of the *ulayat laot* are mostly constructed in an unwritten form, sourced from local wisdom. The term '*ulayat*' in Indonesian legal system, has long been known in the Indonesian history of law [28]. However, in reality of the centralization of law in Indonesia has been narrowed the position and the states perspective on *ulayat* right. The transfer of right is often done in a manner inconsistent with the *adat* law. The transfer of rights that does not heed *ulayat* rights will be opposed by the community. The transfers that take place will cause a clash to a bloody conflict between the licence holder and the public [29, 25].

In the current context, the rights of the *ulayat* themselves are in accordance with the concept related to the rights to manage, in which the state delegates the authority of state control over right to manage to the regions (province and district) and the indigenous people. However, Law no. 5 of 1960 on the Agrarian laws determines a number of conditions of right to manage of *ulayat* for indigenous people namely; (1) as long as the fact remains; (2) it should be which in accordance with national and state interests, based on national unity; (3) should not conflict with other laws and (4) other higher regulations. This requirement are very difficult to achieve by the *adat* community. This requirement is considered by the *adat* law community as a form of doubt and suspicion of the state against them [5].

The above four coditions also apply to Aceh. All laws relating to land, sea and resources are always associated with the national agrarian law. In such context the rights of *ulayat laot* should also be seen in connection with the Agrarian Law. *Panglima Laot* is an institution related to the *ulayat laot* right.

According to T. Rayeuk, there are a number of principal signs in the institution of *Panglima Laot*, namely [30]:

(a) *Panglima Laot* is the head of the institution of *adat laot* (Customary Law) chosen by the *pawang* (sea master) as the fisherman's representative of the territory. *Panglima Laot* must come from the circle of *pawang* and understands the *adat laot*, because *Panglima Laot* beside as the head of *adat laot*, he also the chairman of the customary court to execute and funish the culprit in fishing comunity

(b) The *Panglima Laot* is an institution who control and implement the *adat laot*. It is not a political institution and must be distinguished from other coastal professional organization such as the Indonesian Fishermen Association (HNSI) as well as other institutions conducting activities within the coastal communities.

(c) The responsibility of *Panglima Laot* to make peace in fishing community (*hukom peujroh*), to protect and maintain the customary law of the sea, to mobilize fishermen to carry out social custom in case of accident or disaster, to lead traditional ceremonies, and to preserve and care for the sustainability of the environment within its jurisdiction.

(d) The institution of *Panglima Laot* exists only in coastal areas and waters where fishermen engage in fishing activities. Thus, this institution does not exist in areas where there is no coastal area.

(e) The object of the customary law of the sea is the fishermen; both individuals as well as groups engaged in fishing activities those living in the coastal areas. Essentially, the object of *adat laot* covers personal or groups who inhabit the land with no regards for their professional background.

(f) The territory of *Adat laot* is formed not on the basis of a particular administrative territory but has been formed in such a way that it is permanent because it is determined by its geographic state. The territory of *Panglima Laot* had been determined since the existence of the institution of *adat laot*. Thus the territorial boundaries of the *adat laot* only follow the geographical conditions of the sea, and they cannot be changed for administrative purposes and the like.

(g) The history of the Institute of *Panglima Laot* is only known at the *lhok* under *Mukim* teritory. (level, which differs in its position with the tiered *Syahbandar* (harbour master) institution.

Based on the description above, what is considered the territory of the *Adat laot* should be separated from the concept of administrative boundary. Therefore, the division and regulation based on the customary spaces are very clear and strict. Even by following the concept of the customary institution's existence, the customary laws and institutions that regulate the use of each space in the customary territorial space are clear.

Under such a scheme, the area under the *Adat laot* is a part of the customary *ulayat* rights of the communities that should be separated from the general administrative area. According to T. Rayeuk, the *Adat laot* areas are not all coastal and marine areas. This region is only part of the coastal areas where fishermen pull the Aceh traditional nets (*pukat Aceh)*, and most of the waters are where traditional fishermen fish [30, 26].

Based on the *Adat*, the territory under the jurisdiction of the *adat laot,* can be seen through the concept of the coastal space distribution covering the estuary (*kuala*), *teupin* or sea shore), *leuen pukat* or area for net repairing to the open sea. In addition to that, the territory enacted by the customary law of the sea (*hak ulayat laot)* is wider than the territory of the customary law of the sea itself. The *Adat laot*'s territory is related to the community's customary rights *(hak ulayat laot)* of the area. But, the territory of the *Adat laot* also applies to fishermen who come from certain areas who are experiencing customary disputes about fishing. According to the custom, disputes that can be resolved are not limited to civil disputes, but also criminal disputes, but the cause of the dispute is the activity of fishermen in catching fish, not others. However, nowadays, moving from the state law arrangements, disputes that can be resolved are limited to civil disputes, especially those relating to fishing activities.

The wider scope of the conflict cannot be separated from the customary law's perspective on fishing activities. Customary law sees the process of fishing and the territory of the *Adat laot* as the unity of ecosystems that cannot be separated. Since certain criminal behaviour during that time will have implications for the *Adat laot*, it should be resolved by customary law. For example, in the case of fishermen who used certain fishing methods in a particular fishing area that did not practise such methods and in a territory ruled by the *Adat laot* [30, 26].
In general, in the division of space, the *Panglima Laot* Institute has the following tasks: (a) maintaining and supervising the provisions of customary law and the customs of the sea; (b) coordinating and supervising any fishing effort at sea; (c) Resolving disputes between the fishermen or their groups; (d) maintaining and administering the traditional ceremonies involving the sea; (e) maintaining trees on the beach that should not be felled as the fish will move away from the sea (needs to be adapted to local conditions and situations); (f) acting as a liaison between the fishermen and the government and also between one *Panglima Laot* with another *Panglima Laot*; (g) improving the living standards of fishermen/coastal communities.

The sea customs, related to the coastal ecological customs, are bound by three main restrictions, namely: (a) prohibition of cutting/destructing trees on the coast of the sea such as arun/spruce, pandanus, ketapang, mangrove and other trees that grow on the beach; (B) prohibition of bombing, poisoning, inducing anesthesia, electrifying, removing coral reefs, and other materials that could damage the environment and other plants or animals in that habitat; (C) prohibition of fishing other marine animals and plants protected by the laws and regulations.

The above prohibition also directly relates to two aspects of the marine environmental problems: (1) the regulation about the use of fishing gear and the prevention of using fishing gear which damages the environment and the restriction on the distance between a particular catching area to certain fish habitat; (2) prohibiting the use of unfriendly fishing gear, removing waste, disposing of vessel repair materials, and disposing of oil into the coastal areas [33, 34].

The three restrictions are interrelated. The prohibition of cutting or damaging trees will largely determine the natural shoreline so various sediments can be retained. Sand-induced wind movement is also hindered by the trees. In addition to the customary maintenance of the environment, there are social customs in the operational life of the fisherman that broadly determine two things: first, at the time of the destruction of the ship/boat or other fishing device at sea, they give a sign by raising a flag as a sign of distress. Boats that see the signal will immediately come close to provide assistance. Secondly, if any fishermen happen to drown at sea, the entire boats will find the corpse for at least one full day and if there is a boat that comes across any corpses in the sea, the boat is obliged to take and bring the corpse back to the mainland.

Basically, the two customs describe the unification of the social circle of the coastal communities and the "land" community. In the "land" society, every occurrence of a disaster, must be solved or assisted as a group. If a disaster is related to a person who died, then the *fardhu kifayah* (Islamic religious communal responsibility) must first be fulfilled before the society can go about their daily activity.

In addition, there are customs pertaining to drifting goods. In this case, every item (boat, ship etc.) that drifted in the sea and found by a fisherman must be handed over to the local *Panglima Laot* for further stewardship. Included in the category of various drifting goods are items that can be ascertained not owned by the coastal communities in the area; the discovery must be submitted through the *Panglima Laot*.

3 Panglima Laot: From The Past For The Future

Panglima Laot has been known since the time of the Pasai Kingdom, even Marco Polo who visited Pasai in 1292 and settled there for 6 months, was always accompanied by a *Panglima Laot* in order to get a good catch [2]. The *Panglima Laot* institution continued to live and flourish during the period of Aceh's Kingdom from the 17th Century to the present [6]. According to Wilson and Linkie, despite existing for more than seven centuries, *Panglima Laot* is still relevant as a source of local wisdom that still manifests itself in Aceh in the field of marine management [31].

In the span of seven centuries, it appears that a number of dynamics had occurred in the Institution of *Panglima Laot*. An important dynamic as mentioned by Anthony Reid is that in the early period of the Kingdom, *Panglima Laot's* main duty was to collect taxes from the ships that stop at a port and second, mobilised the people especially fishermen to fight in a war. At the time of the last Sultan of Aceh, Sultan Muhammad Daudsyah (1878-1939), Teuku Umar Johan Hero was appointed as the *Panglima Laot* (*amir ul-bahr*) of the Kingdom of Aceh in 1886. On behalf of the *Panglima Laot*, he collected a tax of $0.25 per *pikul* (62.5 kg) for the Sultan who was based in Keumala [1].

In addition, in the last Kanun Syarak Kingdom of Aceh issued in 1270 H (1854 AD), *Panglima Laot* was subjected to the *Hukum Laksamana* (Admiral's Law). According to Snouck Hurgronje, *Panglima Laot* was no longer an extension of the Sultan's authority but governed the customs, marine practices, and related social life in the region [4].

According to the authors, the changes recorded by Hurgronje were very political at the time. As a colonial advisor, such an opinion was considered ambiguous and intended to bypass the influence of *Panglima Laot* in mobilising the war against the colonial power. By simply sticking to the customs, marine matters and social life, *Panglima Laot* no longer strived to be an important force mobilising the power of war.

Van Volleh Hoven as the founder of the Leiden School of Leiden said the position of *Panglima Laot* is also structurally very important. According to Hoven, *Panglima Laot* officially regulated in the state, which governed the fishing areas on the basis of the Sultan's letter. The mention of the Sultan's letter affirmed the fact that *Panglima Laot* too had the authority to rule over its territory on the coast. This was confirmed by Hoesein Djajadiningrat who mentioned Lhok's *Panglima Laot* as the head of a *lhok* or *kuala* (estuary) or bay, who headed a number of trawls [26].

The generations of customary researchers agreed with Teuku Djuned findings. He mentioned that the problem of *Panglima Laot's* power had already taken place in the community of the *Adat laot*, either before or after being regulated by the legislation. It is important to note that the issue of the sea custom is something that is already engrained in the fishing communities in Aceh [21].

The above description illustrates the spirit of the local people in Aceh in sustainably utilising marine resources. This is particularly true with regards to local wisdom; there were also rules and customary courts, as well as having a territory with clear resources.

In addition, the Institutes' functionaries also undertook a revitalization effort for the young generation in Aceh. Referring to the Decision of the Meeting of *Panglima Laot* in Aceh 23-25, 1992 in Langsa (East Aceh), it was revealed that the effort to write, in terms of the effort of the revitalization of the customary law of the sea in Aceh, is a conservation effort that started since 1990. In particular, in cases involving the ethics of various fishing methods, *Panglima Laot* had been implementing deliberations to formulate since January 14, 1972 in Kabupaten, Aceh Besar District.

The effort to create a written customary rules has been made by the Aceh Special District Office since 1978, through the Supplementary Regulation of Sea Customary Law/ Procedures for Fishing with Fishing/Free Fishing in the Free Sea between Trawler Boats and Boats using Paddles/Aceh Trawls and Conclusion of the Deliberation of *Panglima Laot* from the whole of Aceh Besar and Banda Aceh.

The situation is important to be examined so that in the future, the *Adat laots*, as a form of local wisdom, can be as an implementation framework for the management of marine resources in Aceh. This offer is not excessive, considering the local reality in Aceh is very close to the life of the community. It may be said that the pattern is already engrained in society.

It is this wealth that is offered for the future of Aceh. The pattern of local wisdom in the form of the *Adat laot* of the sea can be used as a model in sustainable ocean management in Aceh. The *Adat laot* is carried out by *Panglima Laot*. *Panglima Laot* and the *Adat laot* in Aceh, in addition to making arrangements about fishing, also regulate the prohibition of the destruction of the marine environment, and the existence of prohibition in environmental destruction, on superstition related to the sea on certain days, have implications for the management of the ecosystem [16].

Moving on from the consensus of the Rio Declaration, the realisation about the importance of local wisdom is in harmony with the global will that promotes the concept of sustainable development at the beginning of the 21st. The concept of sustainability has become a global consensus. A total of 177 countries that met for the Rio Declaration had to apply this in various provisions in the country.

The general concept that can be traced from the Rio de Janeiro Declaration is about the capacity of the current situation to not diminish the interests of future generations. One of the reasons that initiated this concept was the concern of the state leadership about the natural degradation of the earth; this was regarded as the failure of the Stockholm Declaration, in particular on the condition of the global environment which had declined in quality after the agreement. Nevertheless, the concept also had produced some successes, among others in the field of normative and institutionalisation of the international world. Along with environmental sustainability, the concept of local wisdom was also discussed. These two issues are related to each other. Environmental sustainability is the ultimate goal, while local wisdom is a pattern for its accomplishment.

4 Harmony Of Marine Resource

In Indonesia, Keraf is one of the intelectuals offering ethics in dealing with the environment and the utilization of resources. One of the principles offered is how the humans should strive to be in harmony with nature. Here, nature doses not always have to be seen as wealth to satisfy human needs. Departing from such concept, the potential of marine resources, is not only seen as a tool to improve material welfare but also maintain the balance of the ecosystems that will contribute to the balance of life in the universe.

This case is determined from the beginning in Law no. 11 of 2006; the authority to manage marine resources should be carried out by taking into account the principles of sustainable development and environmental conservation. Another emphasis is in conjunction with the maintenance of customary law in Aceh.

There are two important points mentioned above, namely the positioning of *Adat laot* and the management of the sea by considering the principle of sustainable development. The core of the *Adat laot* which deals with environmental issues, does not escape the context of sustainable development. The above rules and restrictions illustrate the goals of sustainable development [13]. By understanding the purpose of sustainability, the great potential of marine resources does not create a policy that exceedingly benefits the current generation. The utilisation process must coexist with the environmentally friendly concept, non-destructive patterns, and always pay attention to future generations.

Aceh has a great potential, with 119 islands and a coastline of 2,666.27 km. The water area reaches 295,370 km2. Its potential resources can reach up 1.8 million tons per year. Seen from the perspective of utilisation, material welfare should be a concern. In fact, until now, the poverty rate of the coastal communities is very high, with a figure of over 20%. However, it is not that way, the problem does not stop at the level of utilisation, but rather it extends to how the treasure of the sea can become part of human life.

Poverty is still considered a factor that causes environmental damage. The solution to the problem of poverty is also one of the targets that need to be globally resolved. Hence, it is a key role of human, individuals and institutions, to determine the fate and future of marine resources. This is partly because the management of the sea should encourage the preservation of the quality, diversity and availability of sufficient resources for the present and future generations, in the context of the fulfilment of food needs for the people, poverty reduction and sustainable development.

5 Conclusion

The potential of the sea utilised in such a manner in order to achieve sustainability must be protected not only by strict regulation from destructive behaviour but also in favour of the local wisdom of *Adat laot* and *Panglima Laot*. Local people view environmental sustainability as something that integrates nature with values.

References

1. A. Reid. Asal Mula Konflik Aceh (Jakarta, Yayasan Obor, 2005).
2. A. Ricci, The Travels of Marco Polo the Venetian, London: Routledge, 1994.
3. A. S. Keraf. Etika Lingkungan (Kompas, Jakarta, 2010)
4. A. S. Usman. Nilai Sastera Ketatanegaraan dan UU dalam Kanun Syarak Kerajaan Aceh dan Bustanul Salatin, (Penerbit UKM, Bangi, 2005).
5. A. Sodiki, Politik Hukum Agraria (Jakarta, Penerbit Konstitusi Press, 2013).
6. Adrianto L, Amin MA, Solihin A, Hartoto DI. Konstruksi Lokal Pengelolaan Sumberdaya Perikanan di Indonesia (Bogor, IPB Press, 2011).
7. Busilacchi, Russ GR, Williams JA, Sutton SG, Begg GA. The Role of Subsistence Fishing in the Hybrid Economy of an Indigenous Community. Journal of Marine Policy **37** (2013).
8. Chaliluddin, A. Purbayanto, D. R. Monintja, M. Imron, J. Santoso. Institution of Panglima Laot in Supporting Sustainable Capture Fisheries Based on Local Wisdom in Aceh Jaya District. International Journal of Sciences: Basic and Applied Research, **16**, 2 (2014).
9. D. S. Adhuri, Saat Sebuah Desa Dibakar Menjadi Abu: Hak Ulayat Laut dan Konflik Antar Kelompok di Pulau Kei Besar, Antropologi Indonesia, **57** (1998).
10. Gadgil, M., F. Berkes, C. Folke. Indigenous Knowledge for Biodiversity Conservation. Ambio **22** (1993).
11. Hardin, G. Tragedy of the Commons, Science 162 (1968).
12. M. A. Abdullah, T. M. Mansur, S. Tripa, Kearifan Lokal di Laut Aceh (Banda Aceh, Panglima Laot, 2010).
13. M. A. Abdullah, S. Tripa, T. M. Mansur. Selama Kearifan adalah Kekayaan (Jakarta, Panglima Laot dan Yayasan Kehati, 2006).
14. Markus T. Towards Sustainable Fisheries Subsidies: Entering A New Round of Reform Under the Common Fisheries Policy. Journal of Marine Policy, **34** (2010).
15. M. H. Daudy, Kedaulatan Laut Berbasis Sumberdaya Lokal: Perspektif Hak Asasi Manusia, Kanun Jurnal Ilmu Hukum, **63** (2014).

16. M. H. Nya' Pha, "Panglima Laot: Peranannya dalam Lembaga Adat Laot", Makalah Duek Pakat Panglima Laot se-Aceh di Sabang, 19-20 March 2001.
17. Ostrom, A General Framework for Analyzing Sustainability of Social-Ecological Systems. Science, 325 (2009).
18. Peterson, N., B. Rigsby (ed.) Customary Marine Tenure in Australia (Australia, University of Sydney, 1998).
19. Pita C, Pierce GJ, Theodossiou I. Stakeholders Participation in the Fisheries Management Decision-Making Process: Fishers' Perceptions of Participation. Journal of Marine Policy **34** (2010).
20. Ruddle K., Akimich, T. (eds), Maritime Institutions in the Western Pacific (Osaka, National Museum of Etnology, 1984).
21. S. M. Syaref,. Leuen Pukat dan Panglima Laot dalam Kehidupan Nelayan di Aceh (Jakarta, Yayasan Rumpun Bambu dan CSSP, 2003).
22. S. W. Ananingsih. The Developments of Customary Marine Rights: A Case Study in Kei Islands, Southeast Maluku Regency. Journal of Wetlands Environmental Management, **2**, 2 (2014).
23. Sulaiman. Pembangunan Hukum Perikanan Berkelanjutan Berbasis Kearifan Lokal. Kanun Jurnal Ilmu Hukum, **16**, 3 (2014).
24. Sulaiman. Kearifan Tradisional dalam Pengelolaan Sumber Daya Perikanan di Aceh pada Era Otonomi Khusus. Jurnal Dinamika Hukum **11**, 2 (2011).
25. Sulaiman, S. Bahri, M. A. Abdullah. Sisi Lain Ulayat Laut: Perspektif Hukum Pengelolaan Pesisir Berbasis Hukum Adat Laut (Banda Aceh, Bandar, 2017).
26. T. Djuned, "Membedah Adat dan Hukom Masyarakat Aceh", dalam L. Munir (Ed.), Menuju Revitalisasi Hukom dan Adat Aceh (Banda Aceh, YRBI dan CSSP, 2003).
27. T. Djuned, "Kedudukan Panglima Laot dalam Hukum Positif di Indonesia", Makalah Duek Pakat Panglima Laot se-Aceh di Sabang, 19-20 Maret 2001.
28. T. I. Elhakimy. "Tata Cara Primair Peralihan Hak Atas Tanah Ulayat Menurut Hukum Adat di Aceh", dalam L. Munir (Ed.). Menuju Revitalisasi Hukom dan Adat Aceh (Banda Aceh,: YRBI, 2003).
29. T. I. Elhakimy, Beberapa Segi Hukum Adat tentang Tanah Pedesaan Aceh, Laporan Penelitian (Banda Aceh, Lembaga Penelitian Universitas Syiah Kuala, 1981).
30. T. Rayeuk, "Panglima Laot dan Wilayah Hukum Adat Lautnya", dalam L. Munir (Ed.), Menuju Revitalisasi Hukom dan Adat Aceh (Banda Aceh, Yayasan Rumpun Bambu, 2003).
31. Wilson, C., & Linkie, M. The Panglima Laot of Aceh: A Case Study in Large-Scale Community-Based Marine Management after the 2004 Indian Ocean Tsunami. Oryx, **46**, 4 (2012).
32. Y. Rinaldi, D. Suhendra, C. Desyana, R. Setiyaningrum. The Document of Analysis on Policies the Management of Fishery and Marine Resources of Nanggroe Aceh Darussalam Province, (Banda Aceh, WWF-Indonesia, Aceh Program Office, 2007).
33. Z. Imran, M. Yamao, "Konstruksi Peran Panglima Laot Lhok Menuju Tata Kelola Kawasan Konservasi Perairan Daerah Berbasis Sosial-Ekologi Sistem", dalam Sulaiman (Ed.), Aceh, Kebudayaan Tepi Laut dan Pembangunan (Banda Aceh: Pusat Studi Hukom Adat Laot dan Kebijakan Perikanan Universitas Syiah Kuala, 2014).
34. Z. Imran, M. Yamao. Toward Strengthening Social Resilience: A Case Study on Recovery of Capture Fisheries after Asia's Tsunami in Aceh, Indonesia, World Academy of Science, Engineering and Technology International Journal of Environmental, Chemical, Ecological, Geological and Geophysical Engineering, **6**, 9 (2012).

PRESERVING SENSE OF PLACE AT HISTORIC WATERFRONTS IN MALAYSIA

Badaruddin Mohamed[1] and *Nurbaidura* Salim[2]

[1]School of Housing, Building and Planning, Universiti Sains Malaysia, 11800 Penang
[2]Sustainable Tourism Research Cluster, Universiti Sains Malaysia, 11800 Penang

Abstract. Historical preservation is important to protect significant historic resources from destruction and inappropriate changes. Urban waterfronts are unique cultural resources that have its own identity, represent the lifestyle of surrounding communities. Over the years, the activities along the waterfront have changed along with city development, resulting in the loss of sense of place. In recent years, the agenda of preservation the cultural value of waterfront areas has been a research focus among the civil society and NGOs. However, in the midst of rapid development, this aspect is often overlooked and deteriorates the historical values of the place. At the same time, the quality of life of waterfront communities is also affected. This conceptual paper is a call for a set of actions to ensure that historical values at these places are protected. Through content analysis from previous literature reviews, this paper investigates how regeneration of historic waterfronts can help in preserving sense of place and improve the quality of life among waterfront communities. Findings from the literature revealed that contextual integration is important for successful redevelopment of historic waterfronts in order to preserve their sense of place.

1 Introduction

The place identity expresses a harmonious balance between people and urban events that are linked to one another, making a place unique and attached to the people [1]. At the same time, these places are moving forward with the rapid development and globalization, contributing to the change of places in urban areas. This situation has resulted to the increasing urban identity crisis, where cities, places, customs and behavourial patterns resemble to one another (Bentley in Sepe, 2013). De-territorialisation of waterfront areas due to globalization and commercialization is a common trend, place identity at the waterfront areas slowly deteriorate. Abandonment of waterfront due to globalization and technology changes has been a focus of research in recent years. As a result, urban waterfront has become derelict, loss sense of place and eventually the community has turned its back. This situation can be seen in many waterfront cities in Malaysia. Kangar Waterfront is one fine example of this situation. Located at the center of Perlis and among recreation nodes of Perlis River, research conducted by Abu Bakar, Ujang [2] revealed that the waterfront fail to function as social and recreational public place due to its degrading condition. Poor water quality, erosion, flooding, vandalism and pollution are some the issues observed at the waterfront, and eventually affected the way community use and engage with the waterfront.

Another good example can be derived from the Clan Jetties in Penang. Although the Clan Jetties is considered as 'living heritage' by the Penang State Government, tourist arrivals seem to be flocked at the clan village. From the positive sides, the local people perceived this scenario as money-oriented business and lucrative source of income to serve tourists' needs. While at the same time, the identity of the place may be deteriorated and forgotten if awareness to preserve sense of place and place identity are not emphasize. Some scholars argued that the loss of sense of place and its meaning contributed to negative impacts, not only to the individuals but also community identity [3]. Although today the Clan Jetty is still home to many people living in this area, many have changed their social and cultural historic.

Realizing that historic waterfront a have strong and unique contribution to the history of Penang, current research suggests the needs to preserve its social and cultural history [4]. This paper will adopt phenomenological reading for urban waterfront areas and suggest factors contributing to sense of place at waterfront areas. In order to achieve this aim, this paper will define the concept of sense of place, its importance and relationship to historic preservation, communities and culture of places.

2 Creating A Sense of Place

Due to its interdisciplinary nature, there is a lack of understanding of 'sense of place'. Although there is unclear definition of this concept, most scholars agree that the 'sense of place' are likely related to: (1) emotional attachment of a place, and (2) interpretive perspective on places [4]. It a multidimensional concept that defines identity (relationship between people and place), attachment (emotional relationship to the place) and dependence (degree to which the place is believed to strengthen behaviour) [5]. Based on the understanding of these definitions, it can be described that physical environment plays an important role in reflecting the identity of its users. While the places influence self-identity, people tend to learn, understand, create and maintain their physical surroundings which reflect themselves. A sense of place is a combination of characteristics that makes a place unique and special. Steel in Hussain, Mohd Yunos [6] suggested three main attributes to shape the sense of place, which includes physical setting, social activities, meaning and image.

In understanding characteristics of a place, studies by previous scholars revealed the importance of attributes that create the sense of place. A sense of place is about creating a place to become unique and special for local community as well as visitors. It should be based on the understanding of cultural identity, social activity, physical environment and heritage of a place. A strong sense of place is a place where people would like to be part of and feel connected to the place. One of the key attributes of a successful place is the ability of the place to encourage social interaction among people. It provides people with a sense of belonging to the environment.

In recent years, the role of cultural and history has become a driving factor in the process of waterfront regeneration. According to Shamsuddin, Sulaiman [7], historic waterfronts are usually redeveloped as maritime conservation and reuse of adaptive heritage buildings. Such development is believed to boost economic to municipalities, help in revitalise the areas and enhance local businesses. Historic preservation commences when people value their heritage. Historic preservation not only adds values to the host communities and visitors, but also provides opportunity for future generations to learn and appreciate the legacy of previous history. Based on the final report of 2003 Michigan Land Use Leadership Council (MLULC), a sense of place can be created with collaborations between government and private sectors, working together to create and maintain liveable urban areas. 'Liveable' urban areas in this context, emphasize on the places where people meet, live, work, shop, invest and grow a business, learn and recreate. In order to achieve liveable urban areas, developing 'green infrastructure' within core urban areas is encouraged. In this context, waterfront redevelopment should have consider development that related to heritage, local people, cultural, recreational, commercial, educational and environment waterfront [7].

One of the examples to create a strong sense of place at waterfront areas is to implement adaptive reuse concept of the surrounding historic buildings. Besides that, community involvement is also important to create a strong place identity so that stakeholders can identify their specific needs and concerns. The means of engaging public participation is to achieve ongoing feedbacks and creative solutions to the recurring problems.

The use of waterfront vacant infrastructures by cultural artists can create a new and different appreciation of the built environment, creating a new product for cultural consumption. For example, the initiative taken by Penang State Government, Malaysia to introduce Penang Street Art around the Georgetown city has rejuvenated the living heritage within the World Heritage City. The 'Marking Georgetown' initiative intended to explore the ideas in art for public spaces in the city. More than 50 arts made from iron wrought sculptures were erected on the selected city's building walls. These arts indirectly help the tourists to understand and see the happenings, culture and how things were decades ago. These street arts also have attracted many tourists to visit the Heritage City for the past few years. Besides help to educate the tourists on the local culture, they also may enjoy to snap pictures with mural. All paintings and street arts at the selected walls tell its own story of the previous local daily routine. There were some of the culture and routine that is still being practice until today. For instance, the most painting in Georgetown World Heritage City is the 'Two Children on A Bicycle' which located at Armenian Street. The painting depicted the time when bicycle was the main mode of transportation in Georgetown (Figure 2.3). Another well known street art is 'Jimmy Choo' at Lebuh Leith where this is the place where the famous designer started his shoes apprenticeship (Figure 2.1).

3 Research Methodology

This research will consider Georgetown Waterfront, Penang as a study area, which covers areas from the Clan Jetties to Fort Cornwallis. Georgetown Waterfront is chosen as my study area due to the rapid changes experienced by waterfront and its surrounding areas. With recognition as a 'World Heritage Site' (WHS) by UNESCO in 7 July 2008, tourism development in Penang has rocketed and improve the economic development of city.

This study will employ qualitative approach by using extensive reviews of urban planning documents in Special Area Plan of Georgetown, Penang. According to Conzen [8], analysis of morphology can be described in three stages which include surveying, mapping and analytical techniques. Xie and Gu [9] employed this method in their research of waterfront revitalization in New Zealand. In this research, data collection will be conducted in three stages. Early phase of data collection will involve a review of governmental documents which includes National Physical Plan, Penang Local Plan, Special Area Plan of Georgetown and base maps of the study area. The documents outlining waterfront development plan will be compared with the actual historic-morphological shifts experienced by the waterfront in recent decades. The documents include the publication by the Penang City Council (*Majlis Bandaraya Pulau Pinang*), Special Area Plan of Georgetown established in 2016 to study the revitalization of the waterfront near the city centre. In addition, Penang Local Plan will also be referred to oversee the study the direction of long term development strategies. Historical documents about Georgetown will be examined to understand the morphological patterns of Georgetown Waterfront.

The second stage of this research will record any changes in urban tourism planning on streets and buildings adjacent to the waterfront area. At this stage, direct observation will be employed. The research will adopt a phenomenological approach as a method of study. Subsequently, the phenomenological approach in this research would be reading and observe the place, derived from Francis Violich's phenomenological reading. An inventory consists of building use will be recorded and plotted to identify variation of urban forms. The inventory tools will also assess past and present use of the heritage buildings at the study area. This process is important to evaluate how changes in tourism development have affected the facade and use of the adjacent buildings. In order to support the data, an in-depth interview with the Penang City Council, PLANMalaysia@Penang (formerly known as Penang Town and Country Planning Department) and local community will be held to identify further information pertaining to conservation, tourism and revitalization issues at the study area.

4 Conclusion

The purpose of this paper is to investigate the importance of sense of place at historic waterfronts. Preservation of historic buildings and adaptive use of old buildings can help in creating sense of place. At the same time, preserve the local identity at the waterfront. Participation from local people is important since they are the ones that have close attachment to the waterfront. Smart land use planning and community development initiative can help in recreating a sense of place at waterfront areas.

5 Acknowledgement

The funding for this project was made possible through the research grant obtained from the Ministry of Higher Education, Malaysia under the Transdisciplinary Research Grant Scheme 2016 [TRGS grant no: 203.PPBGN.67611002].

6 References

1. Sepe, M., *Planning and Place in the City: Mapping Place Identity*. 2013: Taylor & Francis.
2. Abu Bakar, E.A., N. Ujang, and F. Abdul Aziz, *PLACE ATTACHMENT TOWARDS WATERFRONT IN KANGAR, PERLIS, MALAYSIA*. Alam Cipta, 2016. **9**(2): p. 33-44.

3. Karsono, B., D. Deni, and C.A. Fithri, *Assessment of Functional and Emotional Attachment in Malacca Riverfront Promenade.* Jurnal Teknologi, 2016. **1**(1).
4. Ng, V., *Toward A Holistic Understanding of Sense of Place: A Phenomenological Reading of Chew Jetty, Penang.* International Journal of Humanities and Social Science, 2013. **3**(20): p. 75-83.
5. Jorgensen, B.S. and R.C. Stedman, *SENSE OF PLACE AS AN ATTITUDE: LAKESHORE OWNERS ATTITUDES TOWARD THEIR PROPERTIES.* Journal of Environmental Psychology, 2001. **21**(3): p. 233-248.
6. Hussain, M.A., et al., *Interaction on Impression of Place at Historic Waterfront.* RESEARCH JOURNAL OF FISHERIES AND HYDROBIOLOG, 2015. **10**(4): p. 112-114.
7. Shamsuddin, S., et al., *Regeneration of the Historica Waterfront of World Heritage Sites in Malaysia: The case of Georgetown and Melaka*, in *UK Ireland Planning Research Conference*. 2010: United Kingdom.
8. Conzen, M.R.G., *Alnwick, Northumberland: A Study in Town-plan Analysis.* 1969: Institute of British Geographers.
9. Xie, P.F. and K. Gu, *The changing urban morphology: Waterfront redevelopment and event tourism in New Zealand.* Tourism Management Perspectives, 2015. **15**: p. 105-114.

Eurocentrism and the Historical Perception About the Malays

Azmi Arifin[1]

[1] Senior Lecturer, History Section, School of Humanities, Universiti Sains Malaysia, Penang Malaysia.

Abstract. This study examines the effects of Eurocentrism to the view of the character, nature and image of the Malays especially in the period of British colonialism in the 19th century. This research explains why Eurocentrism creates confusing and negative views of the nature, civilisation and wisdom of the indigenous people. Eurocentrism is a form of thoughts that often measures and defines non-European civilisation through historical, cultural, religious, geographical, scientific and progressive perspectives based on Western or European values, which is considered to be supreme. The beliefs transpired in the minds of scholars and European politicians as early as the 16th century, and some were extended in the writings about local history by means of the roles played by the European colonial explorers and administrators. In the process of colonising and exploiting the wealth of the indigenous people, the European explorers and administrators often created records that could explain not only the economic potentials of the area that they intend to colonise, but also depicted the inferior characteristics and civilisations of that society to justify their colonisation attempts. In the process, they produced a record of colonial history which later became key references to historians to understand and explain the traits of the indigenous peoples. The ideas behind that colonial history were not only biased as the result of exaggerating the ignorance and retrogression of the indigenous people; it even sought to undermine the wisdom and civilisation of the indigenous people. This is to highlight their own superiority and noble values when placed side by side with the values of the indigenous people. Despite the prejudice, the Eurocentric colonial history is very influential in Malaysia's historiography until this day. This study explains how the influence of thoughts has blurred the understanding of the actual traits of the indigenous peoples and create an apparent confusion over the history of Malay society.

1 Introduction

This writing examines a form of thought, belief or prejudice known as Eurocentrism that strongly influences the historical interpretations of the Malays until now. The main purpose of this study is to explain how embracing these thoughts and prejudices have dictated the historical interpretations of the characteristics, practice and the political knowledge of the Malay society before and under the British colonialism. The study focuses on the historiographical traits of political institutions and the pattern of governance in the Malay states within the Malay Peninsula before the arrival of British colonialization which began with the Pangkor Treaty in 1874.

The influence of Eurocentrism on Malay politics is rarely discussed. Therefore, many scholars are unaware of their existence or understand its true influence on the interpretation of Malay political history about the Malay states. Instead, local historians, until now, are generally more inclined to take on any concept that the Europeans have introduced to explain the early political nature of the Malays. Full reliance on the European concepts and colonial references cause the explanation given about the traits and practices of the indigenous community to be misrepresented. They are often not based on the actual knowledge and the true nature of the indigenous society, but rather as a reflection or comparison of the differences in values between the Malay and European civilisation. In reality, the images and concepts brought into Malay political historiography are more of a picture based on the perception and prejudiced thoughts of the Europeans about the Malays. They are not a comprehensive and empirical study of the true nature of the indigenous community itself. As a result, the historiography that deals with the political

[1] Corresponding author: azmiarifin@usm.my

characteristics of the indigenous people today is not an entirely true representation of the indigenous history, but rather a historical recollection seen from the eyes of the Europeans (Tregonning, 1958; Goh, 2007).

2 Eurocentrism in Historical Thoughts

According to Blaut (1993:1), *Eurocentrism* or *Eurocentric diffusionism* is a form of thought that strongly believes that Europe is a centre of civilisation excellence and it drives human progress either in specific locations or around the world. This theory or thinking explains how progress or civilisation moves and evolves from one centre to another. The ultimate centre of civilisation is considered the reason for the spread of civilisation to other parts of the world that are considered far more backwards and inferior. In this context, Europe is seen as a centre that triggers, propagates and develops all other civilisations. Therefore, it operates as a historical creator, while non-Europe only makes progress as an imitator and recipient of the European civilisation (Blaut, 1993: 1).

> Europeans are seen as the "makers of history". Europe eternally advances, progresses, modernize. The rest of the world advances more sluggishly, or stagnates: it is a "traditional society". Therefore, the world has a permanent geographical center and a permanent periphery: an Inside and an Outside. Inside leads, Outside lags. Inside innovates, Outsides Imitates ... Europe, eternally, is Inside. Non-Europe is Outside. Europe is the source of most diffusions; Non-Europe is the recipient.

Blaut emphasized the difference between the true nature of European progress and myths, assumptions and perceptions, deliberately designed and exaggerated, aimed at creating the wrong view about European excellence against the inferiority of others. These myth, assumptions and perceptions are formed without objective, scientific or evidence-based research. According to Blaut (2000:4):

> I use Eurocentrism to indicate *false* claims by Europeans that their society or region is, or was in the past, or always has been and always will be, superior to other societies or regions. The key word here is "false". It is not Eurocentric to prefer European music to other music, or European cuisine to another cuisine. It is Eurocentric to make the claim that Europeans are more inventive, innovative, progressive, noble, courageous, and so on than every other group of people; Or that Europe as a place as a more healthy, productive, stimulating environment than other places. It is not Eurocentric to extol "England's green and pleasant land", but it is Eurocentric to claim that this land is greener and more pleasant than all other lands of the world.

According to him, Eurocentrism is a form of European-focused thinking, which evaluates and makes a comparison between a non-European civilisation with Europe, based on false assumptions. It even aims at creating contempt, by trying to show Europe as more noble, superior, intelligent, powerful, advanced, astute, and supreme compared to non-Europeans. For example, the European political system which is a form of democracy, is regarded as the ultimate political system in contrast to the political system of the non-Europeans, such as those based on Hinduism or Islam, or the climate and physical geography of Europe are the best compared to the climate and the physical lay-out of Asia, or the Europeans as humans are far more civilized and distinguished compared to the indigenous people of Africa, India or South America, or the European culture is more noble and outstanding than non-European culture (Amin, 1989; Quijano and Ennis, 2000: 533-80; Lander, 2000: 519-32; Alatas, 2001: 49-67; S. Sayyid, 2007: 300-16; Hobson, 2009: 217-38; Forstater. 2009: 63-76; Majid Amini, 2010: 29-45;).

The idea that elevates Europe as more noble and superior than non-Europeans emerged in the thoughts of Europeans since 16th immemorial, but it became so significant at the end of the 19th century. This is because such thoughts have been driven by the emergence of the ideology of colonialism, racism and the theory of social evolution that grew rapidly in Europe, particularly in Britain during that period. During that period, there were two important developments that strengthened Eurocentrism (Azmi Arifin, 2007; Chapter 2). First, the emergence of the social evolution theory and theory of progress (the idea of progress) brought about by well-known European thinkers such as Montesquieu, Voltaire, Hegel, Marx, Spencer and Darwin. They claimed that the level of human progress and change could only be measured through the hierarchy of progress accomplished by Europe, and cultures that are far apart from it (to be called primitive or Asiatic) are the most inferior civilisation. Therefore, the European thinkers deemed some of the non-

European worlds as the missing link to the point that they equate Africans and Southeast Asians as apes, chimpanzee or Orang Utan (Steadman, 1969; Marshall and Williams, 1982; Savage, 1984; Miles, 1989).

Second, the development of imperialism witnessed the opening of many new colonies and the widespread expansions of new myths about the violence and the inferiority of the Asian civilisation. Reports submitted by colonial explorers and administrators about despotism and feudalism in Southeast Asia, for example, have been extensively misrepresented as justification to reinforce the European perceptions about their advancement and the decline of non-European societies, which were then considered to experience the stages of stagnation in European history. Therefore, the uneducated and backwards Asian people who were equated to slaves seen in the days of European feudalism who deserved to be treated with contempt in the name of progress and enlightenment (Savage, 1984).

The political and economic domination of the non-European world through the success of military and colonial expeditions had created a perception of Europe's greatness vis-à-vis the inferiority and retrogression of the non-Europeans. In the process of distinguishing themselves from the colonised natives, the Europeans such as the British, often make notes of their superiority and their virtuousness so a comparison could be made about the inferiority of the indigenous people whom they called Asiatic-feudalists. Among the notes made included reinterpreting the progress of the indigenous history according to the 'tunnel' or 'hierarchy' of the European historical progress, based on the historical phases experienced by the Europeans. This situation had influenced the formation of a historical interpretation grounded in the theory of social evolution; a belief that since the European civilization is the best, then the whole non-European world needs to go through the same historical stages as those experienced by the Europeans in order to make progress, with the European being seen as the ultimate race of the highest level or hierarchy. The deduction was often made that Europe had already undergone various historical stages which included the primitive, slavery, feudal and capitalist age, and in the 19th century moving towards the age of transition, also known as the industrial age or modern age. Hence, occupied colonies which had not gone through similar phases were considered backwards and left behind; they were believed to be at the lowest level of the European history, known as the Asiatic, Slave or Feudal times (Azmi, 2007: Chapter 2).

Imperialism in Asia and Southeast Asia had allowed the Europeans to assert their belief in superiority; thus, confidently and fully utilising the ideology of Eurocentrism to whiten their exploitation policies. Indeed, by depicting the indigenous people of Asia and Africa as those from an inferior civilisation, European colonialism was successful in asserting the myth of white man's burden in order to uphold the assumption that non-civilized and non-European natives should be colonized for the sake of civilisation, and that the holy duty was determined by God only for the greatest nation. As such, British colonialism is, for example, regarded as a process of civilisation and enlightenment of non-Europeans as opposed to other forms of colonialism that were driven by the intention to oppress and exploit (e.g. Japanese colonialism). As explained by some researchers (Joseph et al., 1990: 1), Eurocentrism evolved through the process of colonialism and economic exploitations and subsequently creating ideological justifications for the colonisation and domination. In other words, imperialism created and used non-Europeans' retrogression and inferiority as an excuse to assert their will and power not only as conquerors, but also as the architect of progress and civilisation, or Europeanising its colonies. Blaut (1993: 2) defined colonialism as '... must mean, for the Africans, Asians, and Americans, not spoilers and cultural destruction but rather the receipt-by-diffusion of European civilisation: modernization."

3 The History of Malay States And Eurocentrism

Eurocentrism evolved strongly in Europe, especially Britain in the 19th century and it had affected the average British, including most of the soldiers and colonial administrators sent to Malaya. Clearly instigated from British intervention in Malay states especially Perak since the 1870s, it was evident that the British were influenced by various ideologies that developed in Europe at that time. Those ideologies included the theory of social evolution and racism that employed the Eurocentric approach to measure the level of civilisation of the indigenous peoples and regarded their inferiority and the greatness of the British as an excuse to intervene and colonise the Malay states one after another, beginning with the state of Perak at the end of the 19th century. In the process of intervention and 'enlightenment', they had produced various official reports and personal accounts which explained not only the potential wealth of the countries they occupied but also about the weaknesses of the political and government systems in those states, often described as backwards, violent and cruel. Most of the records were lodged by British soldiers and administrators who

had to face the opposition of the Malays especially the Malay ruling elites. As expected, the rulers and the Malay political system were portrayed in an extremely negative manner.
For example, during the intervention in Perak, many British soldiers and administrators had produced records about the Malay political system and the Malay rulers whom they were fighting against, to illustrate the overall 'evil and violent' nature of Malay politics. They claimed that the hostile nature of Malay politics could be found not only in Perak but throughout the Malay states during the 19th century, which was also a true representation of the period before that. They made a comparison of the inferiority of the native political system to the superiority of the European civilisation, with the assumption that the European system was extremely fair, noble and supreme. MacIntyre (1967: 71-72) in his study exemplified various British administrators' reports that suited the nature and purpose of its creation. The reports highlighted a distinct colonial ideology, as well as the elements of that, formed the European views and representations of the natives. Non-European regional political units, including Malaya, in the early period of British colonialism are often seen as 'lawless', 'uncivilised' and 'barbarous'. The typical attitude or views of British colonial administrators regarding the Malays can be explained, for example, from the statement of a British colonial officer in Singapore, 'The innate superiority of the ordinary Englishmen in his sense of honour and justice, is sufficient to dominate the inferior characters of the Malays' or from the words by Hugh Clifford, 'I, the European, the white man, belonging to one of the civilized races in the Old World; The Malays, civilized too, but after the fashion of unchanging Asia, which differs so widely from the restless progressive civilization of the West. '(Quotes by Azmi, 2012: 50-51)

An impression existed of the ultimate state of progress and peace if the Malays, whom were regarded as 'half-children and half-demon', 'semi-civilized' and 'barbaric' are under the domination of the European, especially the great British people. According to Earl of Carnovan, 'It should be our present policy to find and train up some Chief or Chiefs of sufficient capacity and enlightenment to appreciate the advantages of a civilised government and to render some effectual assistance in the government of the country.' The attitude and belief in 'racial superiority' triggered a sense of burden and responsibility among the British colonisers towards the natives to bring them progress and glory. Therefore, Harry Ord expressed the need of a fundamental intervention policy; 'I feel that it would be greatly to the advantage of the settlement if our influence could be thus extended over the Peninsula and I shall not fail to avail myself of any opening that may present itself for doing so.' (Quotes by Azmi, 2012: 51).

As a result of the imperialist thoughts, it was noted that there was a major feature in the colonial reports of the indigenous peoples, namely the dichotomy of the history of the Malay states before and after the British occupation. Prior to the arrival of the British, the Malay states were said to be plagued by terror, feudalism and anarchy. The whole Malay Peninsula, according to Ord, was in the hands of 'the lawless and the turbulent ...', while Frank Swettenham claimed, 'In each state, the ruler whether he was sultan, king, or chief of lower rank, was supreme and absolute. His word was law, and oppression and cruelty were the results.' In the context of the state of Perak before the introduction of British rule, it was said that the government was run using the 'feudal' system, which was barbaric, cruel and despotic Middle Ages. According to William Maxwell, 'The system of government in the various states is despotic. The rulers - whether sultans, rajahs, or what not - have occasionally to fight for their authority ... there was no attempt at a proper administration of justice, simply the strong and wealthy dominated and oppressed the poor and the weak.' The people of Perak before being 'rescued' by the British were claimed to have suffered from slavery, piracy, oppression and anarchy, while its rulers often disputed and fought with each other to seize power. The greedy, wild, savage, and uncivilised traits of the Malay feudal rulers were likened by a British soldier, Captain J. F. McNair as a pirate's behaviour and similar to the bandits on the Rhine River in Germany during the Middle Ages. Under the 'feudal' indigenous rule, it was claimed that there could never be security and safety. What was rampant was oppression, chaos and fear. These situations were the reason to justify the British interference in Perak, as British officials claimed, the British has the responsibility '... to rescue, if possible, these fertile and productive countries from the ruin which must befall them, if the present disorders continue unchecked.' (Quotes by Azmi, 2012: 52-53)

4 The Impact of Eurocentrism On Historical Thoughts Of Malay Politics

The impact of Eurocentrism on the understanding of the Malays or Malay state is very significant. Even the present history of the Malay states to date is in the form of Eurocentric history which is very distinct in nature. The influence of Eurocentrism can be detected through the overwhelming dependence of historians on Western historical concepts that emerged during the colonial times. The borrowed concepts gave a negative picture of the Malay political system because the concepts of the European past were deemed sufficient enough for a retrogressive society and inferior civilisation compared to superior civilisation like Europe. It is not an exaggeration to claim that historians in Malaysia until now are simply interested in borrowing the concepts of the European history without trying to build their own concepts of local history based on the values and the comprehensiveness of the Malay political characteristics. The dependence on European concepts derived from Eurocentrism has created many misconceptions about the Malay political features. These misconceptions have deterred historians and scholars in various other areas from studying, researching and making more concise and accurate conclusions about the nature, the practice and the local political knowledge of the early indigenous peoples.

By borrowing the negative European concepts, scholars in Malaysia have indirectly made comparisons between the inferiority of the indigenous people against the superiority of the European civilisation. For example, researchers have fully utilised the concept of ' rulers with absolute rule' (absolutism) and feudalism to explain the practices and political systems of the locals from the time of Malacca to the 19th-century British colonialism. The concept of absolute power, arising from the historiography of the 18th-century European politics and political rule of government, for example, France under the reign of *Le Roil Soleil* Louis XIV, was used to illustrate the absolute power, the rule of a despotic and cruel Malay government against his subordinates. The absolute monarchy concept, criticised by European thinkers as an oppressive and outdated system, was widely believed had been practised in Malay political traditions since the time of the Malay Sultanate of Malacca. Historians tended to claim that the Malay rulers who ruled Malacca and the Malay states until the 19th century ruled with absolute iron claws such that their tyranny and injustice were not to be questioned by the people who were forced to obediently and blindly obey the rulers.

Similarly, the Malay political system was also claimed to be feudalistic. In this context, the intended feudalism is as mundane and oppressive as the Medieval or Middle Ages system, which saw royal rulers gaining full control of the economy by deploying and enslaving the subordinates; that the king and the ruling classes were two institutions that continuously feud with each other for power, thus creating the state of anarchy, civil war and the collapse of the Malay states. The disunity led to the British intervention who allegedly tried to protect the states or the people from the devastation and violence of 'feudalism' that took place. The historical concepts above, simply taking just a few examples, Asiatic, despotic, absolute monarchy and feudalism, are the creation of the Eurocentric thinking which tried to evaluate the political system and government of the Malay states plagued the backwards and outdated European past practices. It did not emerge from an in-depth study of Malay political practices, but only through assumptions and myths aimed merely at elevating Europe and degrading, even insulting, the Malays. At the same time, the negative European concepts were widely used as a tool to justify British colonialism in the Malay states. For example, the political split and the turmoil caused by 'feudalism' in Perak was the reason used to humanise British colonialism in Perak and other Malay states at the end of the 19th century (Azmi, 2010).

But the colonial concepts had finally been accepted, regarded as true and further developed by local historians who were critical of the Malay political institutions and were influenced by Eurocentrism. Of the local scholars who conveyed the concept of absolute powers included Muhammad Yusoff Hashim and Zainal Abidin Wahid to various scholars from various fields who developed the concept of 'Malay feudalism' - Kassim Ahmad, Syed Hussien Alatas, Chandra Muzafar, Syed Husin Ali and Cheah Boong Kheng - all of them in their respective writings tried to portray the misconducts of the Malay political system as featured in the colonial era. What is most prominent is the historical outlook set by Syed Hussein Alatas, famous for his *The Myth of Lazy Native* book. After strongly criticising the colonial myths about the lazy natives and Raffles' outlook about colonialism that he believed racist in nature (Alatas 1972, 1977), Syed Hussein himself developed the concept of 'Malay feudalism' that was initially introduced in the colonial writings which contained the concept of the lazy indigenous people itself.

Therefore, this brief study wanted to clarify that the concepts and historical interpretations of the early Malay political system and practice are largely, if not entirely, dominated by a form of negative and contemplative Eurocentric thinking or Eurocentrism. Hence, the interpretation of the early Malay political history has to be reshaped and revisited, by reflecting on some of the major shortcomings inherent in today's interpretation, and by taking into account the local knowledge and values that the indigenous people truly held. It is also time for local historians to build their own historical concepts that are consistent with local historical values and far from the Eurocentric influence and the creation of negative images created for the purpose of justifying colonialism. The concept of an absolute ruling monarchy never existed and was unsuitable for explaining the Malay political traits and Malay government, which was never created based on the ruler's absolute power and the absolute loyalty of the people towards its rulers. In the history of the Malays, incidents of rulers or the ruling class being disputed, even murdered, are common phenomena that had occurred since the time of Malacca Sultanate, although it cannot be detected in the European political practice during the same period.

The Malay rulers in the 19th century were also not absolute rulers with wealth and massive soldiers like King John of England or King Louis XIV of France who needed to fight and compete with his nobles to strengthen his rule. The rulers ruled with the support of the ruling class, and the relationships between them was usually very close. Their relationships, as depicted in the texts of traditional Malay historiography, such as in *Misa Melayu*, is likened to a government in an ark. The King would be unable to lead the ark without the cooperation of the ruling class, and therefore, the King had to channel certain powers to his ministers in accordance with his role in a government. The King did not make his own decisions in matters involving government policy, for example in diplomatic or warfare relations, as every decision was taken with the advice and consent of his principal advisors or ministers. This clarified that there was no concept of the absolute royal rule in the Malay states. The leadership of the Malay states as depicted in Malay literature is based on practice or consensus, not based on the enforcement of power and despotic rule. It can also be explained that the values of 'democracy' are not strange in Malay political practice, as opposed to the claimed made by the colonial sources (Andaya, 1975).

The Malay monarchy in Melaka in the 15th century as well as in Perak in the 19th century were not similar to the feudalistic-aristocrats in Western Europe Middle Ages who needed to compete or challenge their king to gain power and liberate themselves. On the other hand, the relationships between the King and the Malay ruling class were very close, they complement each other, and the Malay sultanate was united because the king cannot rule unaided while the ruling class do not have the power or followers. There is no evidence in history that can prove that the ruling class or advisors, driven by strong fortitude that they obtained through their wealth and many followers, were able to challenge a Sultan and compete for his sovereignty even if a weak or poor Sultan reigns. This is because the principle of the relationship between the King and the Malay ruling class was not built based on a feudalistic relationship, but based on a strong bond and relationships between them. This is the reason why a harmonious relationship existed between the government in the palace and his advisors before the intervention of the colonial power.

The close and harmonious relationship between the ruling class explained why governmental monarchy in the Malay states generally ruled well and allowed prosperity to be achieved, thereby preventing widespread oppression and injustice of power. Although it was often stressed that early Malay political institutions were so oppressive and cruel to the subordinates, it was merely assertions and accusations that were deliberately created by the British to justify their intervention in the Malay states in the name of upholding the security and justice in those states. But in reality, the states were in a state of peace, prosperity, and stability until they became fractured and became uncontrolled as a result of foreign interventions that deeply interfered in the Malay political system to create chaos and disorders.

5 Conclusion

In general, it can be said that there are widespread misunderstandings about the political system found in the Malay states before and during the era of British colonialism. This was largely due to the strong influence of Eurocentric thoughts on the idea of imperialism which had encouraged negative records of the history of the Malay political system. The Eurocentric thoughts which became the backbone of the formation of myths about the crimes and the destruction of the native's political system caused by prejudices, rather than scientific studies about inferiority and indigenous crimes that were never supported by clear evidence. The portrayal in the colonial record that the relationship between the rulers and the people in the Malay states was in the form of conflict before the colonial period of the British and the devastating civil war and widespread oppression that ensued were inaccurate and lacked fairness. They were deliberately enlivened and questionable forms of deceptions. Negative European intangible concepts such as absolute and feudal rulers do not exist except in the form of prejudice and bias portrayed by colonial administrators aimed at lowering the Malay political system and establishing the justification for colonialism in the Malay states.

What is clear, the political system and the Malay sultanate institutions have been so systematic, harmonious, effective and capable of meeting the goals of the government and government in accordance with the values, knowledge and needs of the indigenous peoples. Feudalism and the rule of the king with absolute power was a foreign practice that only existed in European political practice but had never appeared to be the actual political practice of the indigenous Malay community as was often claimed. Conflicts, civil wars, massive oppression and the collapse of the Malay state allegedly resulting from 'Malay feudalism' were also myths that had never existed, but intentionally created and exaggerated to justify British intervention in the Malay states which are accused as backwards and oppressed.

Acknowledgment

The author would like to sincerely thank Universiti Sains Malaysia for funding this research through the research university grants (1001 / PHUMANITI / 816232).

References

1. Alatas, S. H. *Thomas Stamford Raffles 1781-1826: Schemer or Reformer?* (London: Angus and Robertson, 1972).
2. Alatas, S.H. *The Myth of the Lazy Native: A Study of the Image of the Malays, Filipinos and Javanese from the 16th to the 20th Century and Its Function in the Ideology of Colonial Capitalism* (London: Frank Cass, 1977).
3. Amin, Samir. *Eurocentrism* (trans. Russell Moore) (New York: Monthly Review Press, 1989).
4. Amini, Majid. 'Allegories of Reason: Eurocentrism and Native Philosophical Resistance'. *Culture, Theory and Critique* 51, 1 (2010).
5. Andaya, B. W. 'The Nature of the State in Eighteenth Century Perak' in Anthony Reid & Lance Castles (eds.). *Pre-Colonial State Systems in Southeast Asia*, Monograph No. 6 (Kuala Lumpur: MBRAS, 1975).
6. Azmi Arifin. 'Feudalisme Melayu: Asal-Usul Pemikiran, Sejarah dan Pensejarahan' ('Malay Feudalism': The Origin of Thought, History and Historiography'. Ph.D Thesis, Department of History, Faculty of Social Science and Humanities, Universiti Kebangsaan Malaysia, Malaysia (2007).
7. Azmi Arifin. 'Perak Disturbances 1871-75: British Colonialism, the Chinese Secret Societies and the Malay Rulers.' *Jebat: Malaysian Journal of History, Politics and Strategic Studies* 39 (2012).
8. Blaut, J. M. *The Colonizer's Model of the World: Geographical Diffusionism and Eurocentric History* (New York: The Guilford Press, 1993).
9. Blaut, J. M. *Eight Eurocentric Historians* (New York: The Guilford Press, 2000).
10. Clifford, Hugh 1875. 'Life in the Malay Peninsula: As it was and is' in Paul H. Kratoska (ed.). *Honourable Intentions: Talks on the British Empire in South-East Asia Delivered at the Royal Colonial Institute 1874-1928* (Singapore: Oxford University Press, 1983).
11. Forstater, M. 'Pan African and Afro-Asian Alternatives (to) and Critiques (of Eurocentrism)' in Kanth, R. K. (ed). *The Challenge of Eurocentrism: Global Perspectives, Policy and Prospects (*New York: Palgrave Macmillan, 2009).
12. Goh, D. P. 'Imperialism and "Medieval" Natives: The Malay Image in Anglo-American Travalogues and Colonialism in Malaya and the Philiphines'. *International Journal of Cultural Studies* 1, 3 (2007).
13. Hobson, J. M. 'What have the Muslims ever done for us? Islamic Origins of Western Civilization' in Kanth, R. K. (ed). *The Challenge of Eurocentrism: Global Perspectives, Policy and Prospects* (New York: Palgrave Macmillan. 2009).
14. Joseph, George G., Reddy, Vasu dan Searle-Chatterjee, Mary 'Eurocentrism in the Social Science.' *Race and Class* 31, 4 (1990).
15. Lander, E. 'Eurocentrism and Colonialism in Latin America Social Thought'. *Nepantla: Views From South* 1, 3 (2000).
16. Marshall, P. J. & Williams, Glyndwr. *The Great Map of Mankind: British Perceptions of the World in the Age of Enlightenment* (London: J. M. Dent & Sons, 1982).
17. McIntyre, W. D. *The Imperial Frontier in the Tropics, 1865-1873* (New York: St. Martins, 1967).
18. Miles, Robert. *Racism* (London: Routledge, 1989).
19. Quijano, A. & Ennis, M. 'Coloniality of Power, Eurocenrism and Latin America.' *Nepantla: Views From South* 1, 3 (2000).
20. S. Sayyid. 'Islam (ism), Eurocentrism and the World Order'. *Defence Studies* 7, 3 (2007).
21. Savage, V. R. *Western Impressions of Nature and Landscape in Southeast Asia* (Singapore: National University of Singapore, 1984).
22. Steadman, J. M. *The Myth of Asia* (London: Macmillan, 1969).
23. Tregonning, K. G. 'A New Approach to Malaysia History: Look at Our Story From Inside, Not From Outside'. *The Straits Times*, 24 November (1958).

Preserving and Promoting Malay Folklores and Legends Using Interactive Motion Comic: The Fang King (Raja Bersiong)

Norfarizah Mohd Bakhir[1], *Mohamad Omar* Bidin[1], and *Ahmad Amirul* Bin Abdul Aziz[1]

[1]School of the Arts, Universiti Sains Malaysia, 11700 Minden, Pulau Pinang, Malaysia

Abstract. The Malay folklore is something that our ancestor inherits to us for so long now. Nowadays, Malaysian folklore is beginning to fade from the face of our country due to the western media that's beginning to dominate the world. Youngsters nowadays prefer to watch western movies and western related stories rather than our own folklore. Not only that, youngsters nowadays is keen towards the western culture due to the innovation of modern technology such as tablets, smartphones, and other devices. There are three important objectives in this research. First is to determine the suitable Malaysian folklore to be converted into an interactive comic. Second is to give a new experience to the audience to enjoy and attract youngsters nowadays. The third is to test the effectiveness of this interactive comic to the young generation from age 13 to 15 years old. This interactive comic is for youngsters who nowadays don't know about the Malaysian folklore and to prevent it from fading from the eyes of our country. As a conclusion, this research is to help preserve and protect our Malaysian folklore from extinction by using modern technologies. **Keywords:** Malay Folklore, Digital Interactive Motion Comic, Kinect

1 Introduction

The Malay folklore is something that our ancestor inherits to us for so long now. It is a treasure that is worth more than anything in our country and it should be preserved for the sake of our ancestors and our cultural identity. Nowadays, Malaysian folklore is beginning to fade from the face of our country due to the western media that's beginning to dominate the world. Youngsters nowadays prefer to watch western movies and western related stories rather than our own folklore. Not only that, youngsters nowadays are keen towards the western culture due to the innovation of modern technology such as tablets, smartphones, and other devices. The question arises from this study is, can the use of technology increase the interest and attention of the younger generations?

There are three important objectives in this research. First is to determine the suitable Malaysian folklore to be converted into an interactive comic. Second is to give a new experience to the audience to enjoy and attract youngsters nowadays. The third is to test the effectiveness of this interactive comic to the young generation from age 13 to 15 years old. This interactive comic is for youngsters who nowadays don't know about the Malaysian folklore and to prevent it from fading from the eyes of our country. As a conclusion, this research is to help preserve and protect our Malaysian folklore from extinction by using modern technologies.
In Malaysia, one of the many treasures that our ancestor hold dear until this day is our many folklores that have been passed down through generations. Malaysia is filled with these amazing stories about local heroes and many other interesting legends that are waiting to be discovered and flourish by our people. Every legends and folklore has its own beauty and moral values for our generation to learn and most of them are daily-life related values that everyone doesn't matter what age they are, they can relate to. The Malay Folklore is a series of traditions, knowledge, and taboo that has been passed down from generation to generation, in the form of word of mouth, written or symbolic. This knowledge is related to the knowledge of Malay ethnic or other related ethnic groups in Malaysia. The understanding of a culture is important to preserve the continuity of the culture so that it won't be forgotten (Ismail Zain, 1978).

Malaysian folklore/cultural heritage is something that our past ancestor passed down from generations to generations. The stories are based on the Malay culture as well as the influence of the surroundings according to the time the stories were written (Othman Puteh, 1984; Md Sidin Ishak, 2005; Hadijah Hamat, 2006). Though sadly our current generation is more attracted to modern films and western fairy tales that the western offered to us. For example, if we ask kids these days who are their favourite hero, their answer will probably be Iron Man, Captain America, Spiderman or even Superman. According to (Dharmamulya, 2008), there are already signs that traditional heritage is becoming more obsolete day by day. Our younger generation does not know about our own local heroes and villains such as Hang Tuah, Raja Bersiong and much more. Thus, our generation is losing interest in our local folklore bit by bit. The content in this old Malay story often includes elements of myths and creatures from other realms relating to Malay culture. There are also narrations that related to the legend in Malay history. The question is, how do exactly maintain or preserve this beautiful tradition?

2 Interactive Tell-Tale Motion Comic

Motion comic or animation comic is a branch of animation that combines fundamentals of comic books. For this project, we are creating a motion comic about the story "The Fanged King" (Raja Bersiong). This research is focused on audience around 12 to 15 years old that is interested in games and comic books. The reason for this method is to promote local folklores and local heritage to the younger audience in a fun and immersive play style. In this motion comic, it will include special effects, sound, dubbing and it will be a story with multiple endings. Motion comic contains dubbing, sound effect, motion, and can even have interactivity. The individual comic panels are animated and expanded by adding sound, voice acting, and animation to each individual panel depending on how the motion comic is drawn. The motion comic is characterized as one of the animated hybrids. Scott Bukatman (2002) points out that online comics are rigid in shape, the characters only move horizontally into a rigid background.

From this project point of view, giving interactivity to the story and letting the audience control the flow of the story will indeed gain the audience attention towards those folklores and legends. With the power of this world current technology and with the use of the mass media, this is something that can be achieved by the power of Microsoft's Kinect Motion Control. The Microsoft Kinect (once codenamed as Project Natal) is a motion sensor developed by Microsoft as a motion detector for the XBOX 360 and XBOX ONE and just newly developed for Windows PC. With the proper usage of this technology, this idea is achievable. Moreover, Kinect will give the audience more control towards the story being told. Interactive system and application are very limited in Malaysia. It is best if one of the systems is based on touch control and motion comic style to act as a platform to gain the younger audiences interest not only children but adult as well.

To give the story more depth and accessibility, the application could be made into a tell-tale game. A tell-tale game is a game that requires the user's decision making to determine the stories plotline. Interactive is something that allows a two-way communication between the user/player and the application itself. Interactive storytelling is quite a different matter. A story would require our demand in the storytelling process; they're supposed to act upon us (Sisonke Msimang, 2017). By adding this feature to the application, the audience has the freedom to control the stories plotline with consequences of their own decision. To put it simply, whatever the audiences choose to decide will give an impact on the storyline of the story. For the aesthetic side of this project, the application will be in a motion comic style of game. It is a new approach to storytelling and this style could be applied to the folklore.

The objectives of this study are to promote to children from age 12 to 15 years old the Malaysian folklore and legend. Second is to help conserve the priceless value of our countries sacred folklores and legends that have been passed down from generations to generations. And third is to introduce a new way of storytelling that is easy for the audience to understand and have full control over the story. The significance of this study is to observe if Interactive Motion Comic using Kinect Touch Control will attract current generation youngsters to learn more about the countries local heritage. By providing options to the player to choose the fate of the character's fate will provide the immersive experience to the user so that they can control the storyline of the story as if they are part of the journey. Hopefully, through this project, they will see our countries local heritage and folklore in a different perspective.

The data were analysed using Brown & Yule (1983) method. The results show that the campaign has been successful in raising awareness about the storytelling of the old Malay story. This topic depicts the whole research. It is about the Interactive Motion Comic Tell-tale Game based on the Malaysian folklore and legends. It explains further about the stories and folklore and even gives the audience full control over the story being told. Other than that, this chapter describes the problem that this issue is currently facing and suggestion how to preserve and conserve the national treasure. In conclusion, by creating an interactive motion comic tell-tale game for the folklores, this will introduce a new way of storytelling and even gaining the audience attention to indulge themselves into the story by letting them control the flow of the story. The application will be in English due to the target audience is for children in Malaysia from age 12 to 15 years old and International audience for the future release. The focus of this project is to introduce a new way of storytelling to the audience.

2.1 The Folklores

Statement from Davis (2010) states that Ben Botkin (1938) has defined the folklore as an item created by a community in the past and it has become an identity for them and this tradition has been passed down from generation to generation. According to Mohd Taib Osman (1988), this traditional storytelling is based on Malay culture that is adapted from the surroundings according to the time the stories were told. However, this traditional storytelling is spread without properly documented. Therefore, the story's content is not the same as the original story.

Malay folklore is stories and legends that have been passed down through generations from our previous ancestors. It is a tradition, knowledge and forbidden things have been passed down from generation to generation in various forms, written or symbolic among ethnic and religious populations around the Southeast Asia region. Malay folklore mostly involves the love stories of royalty, the heroes and the girls in distress. This is because there is still a Malay sultanate in the Southeast Asia region that supplies the content of the story. While this story is choke full of additions and reductions, the noble values and wisdom are not neglected. Therefore, Miller (2013) and Roslina Abu Bakar (2013) point out the importance of maintaining the authenticity of the originality of the stories while Bronner (2012) focuses on the benefits of folklore to the public.

The Malay Folklore is often told by a professional storyteller or a *Penglipur Lara*. The task of this *Penglipur Lara* is to entertain the public with stories of full of noble values and wisdom. Usually, the stories contain comedy, and sometimes all the stories contain *syair, pantun, seloka* and much more. It is one of the essences that give Malay literature. Once upon a time, *Penglipur Lara* is known as the local entertainer that acts as the news informer, entertainment, and knowledge.

Penglipur Lara usually consists of farmers or people from the rural area. They will travel from village to village and bring stories to these places. The authenticity of the stories is still maintained but sometimes adjusted according to current needs and circumstances.

Fig. 1. Interactive Motion Comic - The Fang King.

2.2 The Fang King

For this project, we decided to use the story of The Fang King or *Raja Bersiong*. The Fang King (*Raja Bersiong*) or his real name Ong Maha Perita Deria was a king who once ruled over Kedah in Lembah Bujang. His dictatorship began during the Srivijaya Era. He was the fourth king of the Merong Mahawangsa bloodline. Based on the folklore passed down through generations, the king was someone who was very cruel and feared by the people of his country. Anyone who disobeys his orders and the rule of his country, he will imprison them or either kill them by ordering his guard and soldiers known as "Panglima Raja Bersiong". There are many version of the Fang King's folklore. Each of them has the same reason on how he grows his fangs, but some version of his stories has a different ending and different plotline.

Nowadays, this myths and legends like the Fanged King who has magical powers, no longer exist. The powers of oppression are no longer subject to kings but in the various forms and degrees. The moral of the story is that the ordinary people can have the courage and change the bad situation into a good one. Noble values and wisdom need to be disseminated by various parties rather than just one hero.

3 Methodology

Before conducting this study, a research had been done to 15 respondents age ranging from 12 to 15 years old to find out their knowledge on local folklore and the suitability using motion comic as a medium for this project. The data collected using this method is much clearer, pure, original and trustworthiness is much higher (Pakas Rao, 2004).

The respondents are students ranging from 12 to 15 years old form Sekolah Menengah Kebangsaan Batu Maung. Based on the observation, their knowledge on this topic is very low. Each student is interviewed for this session to gain their knowledge on this topic. As the writer interviewed them, it is proven only a few of them know our local folklores, especially The Fang King. They weren't educated about the magnificent world of our local folklore and they responded that they only watch western and anime movies from the internet, television, and cinemas.

The writer asked if they know these local figures which are *Hang Tuah, Badang, Merong Mahawangsa* and The Fang King. As expected, almost everyone did not have a clue who were they and never heard of them. Three of them knew who Hang Tuah is, but only in name. As for the rest, they did not. For motion comic, some of the students know about it, which they said accidentally stumbled on YouTube. One of them said that they have seen the Watchmen Motion Comic on YouTube and the respondent was really amazed by it. As for the rest, they are all aware of comic books but they do not know about motion comic, so the writer had to show an example of a motion comic and they were amazed.

As for the second phase, the final application of the study is then presented and opens to public testing at Pelinggam Cahaya, Pusat Pengajian Seni, Universiti Sains Malaysia (USM). Within that period, the number of technique that can be used to produce this project is determined and learned. The main reason this project is presented openly in the hall is to obtain data and feedback from users that consist of people of different ages, gender, races, and religion. Based on the observation, Most of them were really excited to play the game and this is a sign that this project might be on the right track.

Though there were quite some negative reviews form the audience. One of them is to add more plotline to the story so that they have the option to choose more decisions and have more multiple endings to the story. This feedback is very helpful to improve the immersive experience of the game. Not only that, the audience suggests that the Kinect sensor should have a more accurate touch control. This is because the Kinect calibration is quite problematic to have the accurate calibration. After discovering these problems, these problems can be fixed in the future for a better experience

4 Summary

The Fang King Interactive Motion Comic is a project that has its own importance and value in order to preserve and promote our own Malay folklore which now is beginning to fade from the eyes of the country. Creating this project was very challenging because it was a new way of telling a story and the story should be based on the true history of the Fang King. During the creation of this project, there were a few weaknesses that this project had to face and mostly the problem is the calibration of the Kinect sensor. Nevertheless, the positive comments from the audience are really a good sign for continuing this project in the near future. Even though there were many challenges and obstacles, this project can be improved and may become a new way of conserving, protecting and promoting our own beautiful Malay folklores.

5 Acknowledgement

This research project was conducted under Universiti Sains Malaysia Long Term Research Grant Scheme, (account number: 203/PTS/6727004). Technical support from the staffs is gratefully acknowledged.

References

1. Bronner, S. J. Practice Theory in Folklore and Folklife Studies. *Journal of the Folklore Society.* **123 (1)**, 23-47 (2012)
2. Brown G. and Yule G. Discourse analysis. *Cambridge: Cambridge University Press.* 288 (1983)
3. Davis, S. G. Ben Botkin's FBI Files. *Journal of American Folklore.* 487, 122 (2010)
4. Dharmamulya Sukirman. Permainan tradisional Jawa. *Bandung. Kepel Press Purwanggani* (2008)
5. Hadijah Hamat. Peranan dan Perkembangan Sastera Kanak-kanak. *Kuala Lumpur: Dewan Bahasa dan Pustaka* (2006)
6. Ismail Zain. Masa Depan Tradisi – dikhususkan kepada pengalaman Kuno di Malaysia 1978. *Journal: Bengkel Wayang Kulit KKBS: 23-26 April 1978* (1978)
7. Md Sidin Ahmad Ishak. Perkembangan Sastera Kanak-kanak di Malaysia. *Shah Alam: Cerdik Publications Sdn. Bhd.* (2005)
8. Miller, S. The County Folk-Lore Series of the Folk-Lore Society. *Journal of the Folklore Society.* 124(3), 327-344 (2013)
9. Mohd Taib Osman. Kebudayaan Melayu dalam beberapa persoalan. *Kuala Lumpur : Dewan bahasa dan Pustaka, Kementerian Pendidikan Malaysia* (1988)
10. Othman Puteh. Sekitar Sastera Kanak-kanak dan Remaja. *Kuala Lumpur: Dewan Bahasa dan Pustaka* (1984)
11. Pakas Rao.P,. Kajian minat membaca bahan kesusateraan Tamil di kalangan pelajar-pelajar tingkatan lima di daerah Petaling, Dissertasi (M.ED). *Universiti Malaya, Kuala Lumpur* (2004)
12. Raja Bersiong. *Retrived from https://en.wikipedia.org/wiki/Raja_Bersiong* (2008)
13. Roslina Abu. Sfera Keindahan dalam Penyampaian Cerita Rakyat Melayu oleh Penglipur Lara Ali Badron, dlm. Zaitul Azma Zainon Hamzah et al., (eds), *Perspektif dan Paradigma Kebahasan*, 228-246 (2013)
14. Bukatman, S. Online Comics and the Reframing of the Moving Image. *In The New Media Book. London: BFI.* 133–143 (2002)
15. Sisonke Msimang,. If a Story Moves You, Act on It. *Retrieved from https://www.ted.com/talks/sisonke_msimang_if_a_story_moves_you_act_on_it* (2017)

Understanding Educational Games Culture in Extending the Relationship between Graphic Appearance and Experiences among Malaysian Lower Secondary

Lilian Lee Shiau Gee[1,1], *Jasni* Dolah[1], and *Norfarizah* Mohd Bakhir[1]

[1]School of the Arts, Universiti Sains Malaysia, 11800, Malaysia

Abstract. Educational games provide a virtual environment for the students, particularly in an endeavour to increase student immersion, and this need of students satisfied with the visual aspect conditions of the game screen to go to the following stage. In determining the effect of visual design on students' learning experiences - it's required a further analysis within the students' motivation, engagement and enjoyment. To achieve this goal, a solid fundamental understanding in theoretical philosophy, which closely related to describe the relationship of visual experience, especially in intrinsic tendencies theory and visual principles that on holds the context of motivation, engagement and enjoyment among students. In this regard, a thorough discussion of relevant theories were led to the suggestion theoretical framework and proposed of Experience Based Design Graphic Style (EBDGS) to identify student's experiences (motivation, engagement and enjoyment) outcomes which mostly depends on the visual design approaches to designing the game scenarios. This paper also aims to describe the theoretical framework proposed in understanding the Malaysian student experience's outcomes by placing the abstract style and realistic style in developing Experience Based Design Graphic Style (EBDGS) game prototype: EBGDS A (Abstract) and EBGDS R (Realistic). The implication of the theoretical framework was focused on the visual design (graphic styles) of games that intentionally centred in the students' experience that tied to the demographic factors (gender, type of schools and areas), and the resulting experiences are examined through the Student Experience in Graphic Measurement Scale (SEIGMS). The intention of this research not only emphasize the suitability visual appearance able better improve student experiences but also increases deeper local knowledge in visual design guidelines in educational game design contexts.

Keywords : Educational Games, Graphic Styles, Students' Experiences

1 Introduction

Lin et al [1] argued that educational game are enacted by students, and the students were also become enacted by the gaming environment, which the design should be constituted with a better visual to enable students to prescribe, sustain and generate personal experiences, onwards increasing continuous in the learning process. In this way, efficiency educational games usage needs overviewed in terms of visual design features that may conducted student's attentiveness [2, 3, 4, 5, 6].

In Malaysia, the implementation of educational games is still in an early stage [7] with concerned the development and evaluation the design context of educational games for enhancing instruction and learning conditions in the classroom. With regards, the preliminary investigation indicating the development of an educational game that comes with the learning approach process needed take into consideration the design context in visual design aspects to arouse the students' interest especially for local Secondary School students. Moreover, most previous studies not centered on

[1] Corresponding author: lilianlsg087@gmail.com

the efficiency educational games usage in the visual design aspect in the suitability of the visual design on student experiences. Additionally, lack of exposure an organized design technique or a style, including less established basic design guidelines to assist designers and educators make an understanding and skills to the concept's suitability of visual design in educational game development on student experiences.

With the lack of empirical studies or incentive information to distinguish the purpose of design (visual aspect) that needs to be adapted in developing educational games in improving the good experience among students.Therefore, this study needs to be carried out to identify how visual design important to shape usability of educational game in enhancing the individual students' affective responses– by taking a closer look at the continually experience outcomes, particularly in terms of motivation, engagement and enjoyment. Thus, this study also explores consistencies intended student experiences by mapping the "visual" details and knowledge aspects of local student culture - as assessment of the degree efficiency educational games towards the ways of appearance. Further, understanding the relationship between graphic appearances and student experiences becomes relevant in an attempt to studying gaming scenarios in accord to factors of demographic that involves the complementary experiential aspects.

2 The Relation of Visual Design and Experiences

In intended that any design for educational games (design-in-use) should concerned with suitability visual appearances (design intentions) by prioritizing students' reactions (experiences), which in relation to local student culture context. This approach the term "graphic styles" as the key analytical concept in discussing and analysing the role of educational games, which verity the relationship between the visual design intentions and local student preferences as closely tend to sustain the local practices (teaching and learning) - also important tries to understand the visual design specifications align with the student's interests, and developing subject-related knowledge in local classroom contexts in relation local culture (demographic factor). In this respect, the demographic factor assumes a close related to student gender, educational area and types of schools.

Correspondingly, the value of visual design to the educational gaming environment is linked to the actual student experiences changes, in relation what kind of graphic styles be able to enhancing and maintaining students experiences and recommendable to developers. Thus, the visual design is valued to be regarded as an effort to understand the hedonic experiences turn in educational game development, which emphasis on students adapt and interact with particular visual representation in relation to their own experience practices and preference criteria.

3 Related Theories

3.1 Motivation Theory

As indicated by Keller [8] "motivational tactics have to support instructional goals", which shows that motivation theory impact on the instruction development and closely linked to maintaining the students' interest in teaching and learning process. Keller was founded four main elements in the learning process that can motivate students or known as ARCS model. The four elements that built the ARCS model stand out Attention, Relevance, Confidence and Satisfaction in growing student's interest in an instructional tool, in relation to understanding student experiences and instruction design.

3.2 Self-Determination Theory

The advances of motivation may restructure an importance experiences impact in educational games. From students' experiences, there is an urgent need to understand the relationship interplay between cognition, motivation and feelings in a process of learning through educational games. In this, the Self-Determination Theory (SDT) becomes the foundation for researchers to acquire, analyse and evaluate the players' experience in games [9]. As a supporter of motivation, Self-Determination Theory implicated three basic psychological needs (competence, autonomy and relatedness) are itemized to identify motivation value in a digital game [10]. Understanding the Self-Determination Theory able to practice assessing the quantity of experiences that could take place in educational games in the appraisal of the significant experiences through the game design features and actors (students) personal complacency.

3.3 Visual Attention

Visual attention actively sustained concentration to a certain number based on the visual object influence the ongoing action and cognition [11]. For enhance a person's ability to recognize certain visual concepts in learning and instruction, visual perception process and experience are intimately related [12]. As added by Chun et al [13], visual attention is an active maintenance of visual processing towards visual items to enhanced ongoing experience. Accordingly, visual attention emphasized knowing individual's experience through the selection of visual features that should attend in learning activities that can clarify the efficiency of visual to delivering knowledge. This also indicates the importance of visualization in a game to attract the concentration of the target audience.

3.4 Gestalt Theory

Gestalt theory describes the significance of visual elements in an object to increase the individual's reception and tendency to attract a person's attention in a certain visual presentation [14]. Instead, this theory function as the suggestion aspect and practicing various visual elements (shape, form and composition) should be presented to achieve efficient visual image results [15]. Thus, Chang and Nesbitt [16], suggested the conception of applying the gestalt principles to assist the visual presentation in games, which is necessary understanding the gestalt principles in producing a visual image in game scene design in order to encourage players' involvement.

3.5 McCloud Theory

McCloud [17] presents a basic theory of icon (a visually image resembling the real object of several aspects). According to McCloud, a continuum from realistic to cartoon images showed increased levels of abstraction iconic. These changes simplify the image of visibility. In addition, McCloud theory distinguishes identification image into three main categories: abstraction, cartoony, and realistic. Thus, understanding the concept of imagery increasing, convenience construct the visual appearances in an educational game.

4 Graphic Styles

The word 'style' in the field of visual art related to the composition and the relationship between the form an overall image, that based on the principle elements, techniques, expression and inspiration [18]. According to Muller [19], graphic style refers as an implementation technique that carried out by artists to portray the mood and features to be conveyed on the audiences. This is simply integrated that graphic display in game are more delineate on surrounding representations and components or objects that are placed in the game environment to enhance gameplay, which the graphic styles can be ranged from abstract to realistic [20]. As this, McLaughlin et al [2] describes the graphic styles are categorized into three different types: i) abstract, ii) stylized and iii) realistic. Each graphic style has been summarizes (see Table 1) as accordance with the form, surface and level of detailed.

Table 1. Description of Graphic Styles (McLaughlin et al, 2010)

Abstract	• Symbolic forms. • Reducing the formal quality • Low in the level of detail • Flat Shadings
Stylized	• Identifiable forms (unrealistic proportions or juxtapositions of parts) • Quality is expressive • The level of detail can be composed from the lower to higher • Shading of curved surfaces or have a texture
Realistic	• Presentation style in photo-realistic modelling accurately the familiar objects. • Level of detail is high. • Realistic shadow casting, Light movement

5 Student Experiences

In addition to theories of learning and visual knowledge, experiences viewed as important facets that is prized for the ability, suitability and efficiency of educational games within the student's interest. In this regards, student experiences will be reviewed in this research were based upon the motivation, engagement and enjoyment to further examining the influence of visual design in educational gaming, which assume a relational meanings and local knowledge aspect of educational game competences by actively engaging the visual design aspect that distributed particular relationships between the student preferences of a visually style gaming.

5.1 Motivation

In games, motivation playing an important function in gaming experiences - as a process of activating, maintaining the player behaviour towards a certain goal, though personal interest and the way player organizing the experiences in gaming [21]. In this way, motivation able pushed the players becoming more engagement and involved with games and turn the games be more successful [22]. With this regard, the relationship between students' experiences with game more domain considering intrinsic motivation to analyse the motivating factors [23]. This means that, the potential usage of educational games need be viewed especially on student experiences in terms of intrinsic motivation.

5.2 Engagement

In games, engagement viewed as the player's degree of involvement in a digital games to attract player's continuous play [24]. Thus, engagement is one of the most frequently cited and needed to consider in digital game for enhancing the student experiences [25]. This indicated that engagement experience is a connector between the students emotional emerged between educational games- that also relies on the educational game design, particularly game features to preserve the degree of interests [26].

5.3 Enjoyment

Enjoyment is in respect of individual's perception of pleasant feeling when conducting an activities [27]. As such, enjoyment regarded as an important source of experience in student learning [28]. Generally, enjoyably viewed as an experience strictly related to motivation and the factors in facilitating the engagement in software application [29]. This resulting enjoyment experience was interlink with motivation and engagement, which need take into consideration in determining the suitability of educational game appearances.

6 Theoretical Framework Overview

In order to identify and analyse the students' experience outcomes throughout interplay between students interaction and visual design contexts, the theoretical framework draws upon with the adoption of Activity Theory (AT) that conducted by Engestrom [30] in 1993. The Activity Theory review the contexts between human (subject) and their reacts upon (object) on a tool that concerned an individual responds, which based on the tools that occur during the reception process of information [31]. With compliments to the theory, the subject replaced with students, the object is substituted with the experiences and the tool replaced with graphic styles - to clarify what experiences result could have within the visually contexts. The Activity Theory system for describing the relationship between graphic styles and students are depicted in Figure 1.

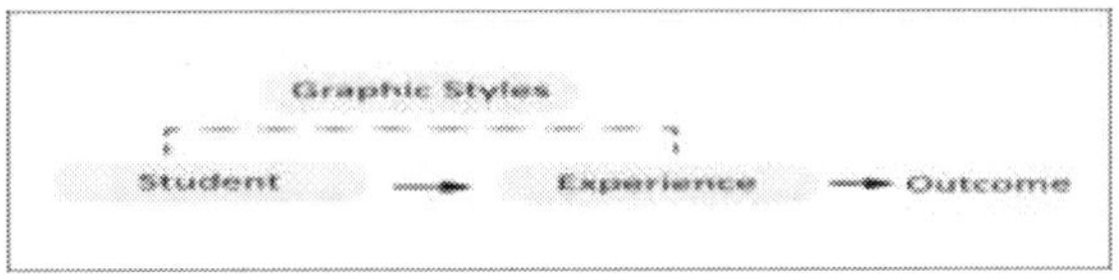

Fig. 1. Structure of Activity Theory

Correspondingly, the theoretical framework was introduced into macro and micro level (see Figure 2). Following the principles of Activity Theory, the macro level tries to indicate the student experiences by viewing students' preference and students demographic as the distinct criteria knowledge for arises local knowledge in game design practices. Simply applied, realistic style and abstract style represent different assumptions about the outcomes of student experiences through different ways of enacting within the demographic factors, which would be making of meaning at the local level of visually context within local student preferences and local syllabus content. Micro level is more concretely in development of Experience Based Design Graphic Style (EBDGS) prototype – a continuum learning activities, game characteristics and visual design represent modes with the intent to understand the student experiences encounter as they emerge through students' enactment of a particular style of educational game scenarios.

This approach to explore the lower secondary students (Form One student) make meaning from educational games in the context of visual display through the ways they enacted and validated their experiences by facilitating and performing within the EBDGS game prototypes: EBDGS A (Abstract) and EBDGS R (Realistic). In this way, Student Experience in Graphic Measurement Scale (SEIGMS) questionnaire are present to generalised the local student view on their motivation, engagement and enjoyment outcomes within the consideration of graphic style quality in designing educational gaming. Thus, students can be seen as "social actors" are free to play and interact with different delivery visual representation modes (abstract or realistic), based on their experience's result outcomes is important to distinguish the substantive quality of visual appearances that imply an interpretive guideline to enhance local student involvements in learning and opportunities for local educational game development.

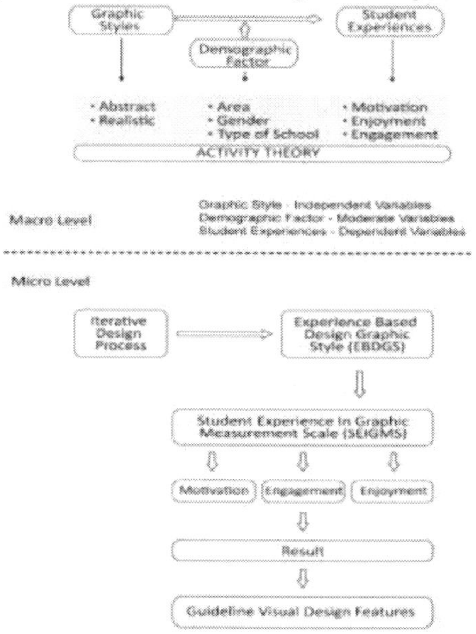

Fig. 2. Theoretical Research Framework

7 Conclusion

This paper explores a theoretical framework for educational game design comprising the graphic styles, demographic factor and students' experiences domain as for understanding the local student perspectives in order to emerge the local knowledge of the educational game development in visual design context. The related theories and key literatures have provided a rich theoretical basis for the theoretical framework development with imply the challenges and issues that inherent incongruities between the game design (which prioritises visual design) in relation to the mode of representation (graphic styles), demographic factor and the students' experiences. In this respect, the relation between particular graphic appearance and students' experiences needs to be determined, for identifying new views on visual design trends that should embedded in educational games, as an external factors that may have influences on local perceptions (students). Hence, the implementation of EBDGS A (Abstract) and EBDGS R (Realistic) is to determine the student experiences by viewing at the graphic style and the experience outcomes of play activities with the aim of understanding the designing concept in increasing the educational game's efficiency without discrediting other current research efforts on understanding educational games. This research assumes that empirical findings were possible capable to accomplish further the local contexts (foreground and background) that tries to formulating the visual design guideline for the local designers and educators to offering an efficient learning presentation and organize a strategic plan enhancing the educational game appearances in advance.

References

1. Lin, Y.L., Lin, H.W. and Yang, Y.T. (2015). Players' Value Structure in Digital Games.
2. Games and Culture, 1-28.
3. McLaughlin, T., Smith, D., and Brown, I.A. (2010). A framework for evidence based visual style development for serious games. Proceedings of the Fifth International Conference on the Foundations of Digital Games, ACM (2010), 132–138.
4. Andersen, E., Liu, Y., Snider, R., Roy, S., Zoran, P. (2011). Placing a Value on Aesthetics in Online Casual Games. In: ACM CHI Conference on Human Factors in Computing Systems.
5. Carroux, L., Le Bigot,L. Vibert,N. (2013). Impact of the motion and visual complexity of the background on players' performance in video game-like displays. Ergonomics, 2013;56(12):1863-76. Doi: 10.1080/00140139.2013.847214. Epub2013 Oct 30.
6. Gerling, K. M., Birk, M., Mandryk, R. L., & Doucette, A. (2013). The Effects of Graphical Fidelity on Player Experience. MindTrek, 229–236.
7. Wang, L., & Lee, K. (2014). The Study Of Learners ' Preference For Visual Complexity On Small Screens Of Mobile Computers Using Neural Networks, Tojet: The Turkish Online Journal Of Educational Technology 13(2), 168–180.
8. Roslina, I. & Azizah, J. (2009). Educational Games (EG) Design Framework: Combination of Game Design, Pedagogy and Content Modeling. 2009 International Conference on Electrical Engineering and Informatics 5-7 August 2009, Selangor, Malaysia.
9. Keller, J. M. (2006). Motivation and performance. In R. A. Reiser & J. V. Dempsey (Eds.), Trends and Issues in Instructional Design and Technology (2 ed., pp. 82-92). Upper Saddle River, NJ: Merrill/Prentice Hall
10. Bouvier, P., Lavoué, E., Sehaba, K., & George, S. (2013). Identifying learner's engagement in learning games: A qualitative approach based on learner's traces of interaction. In O. Foley, M. T. Restivo, J. O. Uhomoibhi, & M. Helfert (Eds.), Proceedings of the 5th International Conference on Computer Supported Education (pp. 339-350). SciTePress: Aachen, Germany.
11. Uysal, A., & Yildirim, I. G. (2016). Self-Determination Theory in Digital Games. Gamer Psychology and Behavior, 123-135.
12. Chun, M. M. (2011). Visual working memory as visual attention sustained internally over time. Neuropsychologia, 49(6), 1407–1409.
13. Silvia, P. J. (2001). Interest and interests: The psychology of constructive capriciousness. Review of General Psychology, 5(3), 270-290.
14. Chun, M. M., Golomb, J. A., & Turk-Browne, N. B. (2011). A taxonomy of external and internal attention. Annual Review of Psychology, 62, 73–101.
15. Koffa, K. (1935): Principles of Gestalt Psychology. London, Routledge and Kegan Paul.
16. Chang, D., Dooley, L., & Tuovinen, J. E. (2002). Gestalt Theory in Visual Screen Design – A New Look at an Old Subject. In Proc. WCCE2001 Australian Topics: Selected Papers from the Seventh World Conference on Computers in Education, Copenhagen, Denmark. CRPIT, 8. McDougall, A., Murnane, J. and Chambers, D., Eds. ACS. 5-12.
17. Chang, D. and Nesbitt, K. (2006). Identifying Commonly-Used Gestalt Principles as a Design Framework for Multi-Sensory Display, Proceedings of IEEE International Conference on SMC, Taipei, Taiwan.
18. McCloud, S. 1994. Understanding Comics: The Invisible Art.
19. Dondis, D.A. A. (1973). Primer of Visual Literacy. MIT Press Design Department. United States of America.

20. Muller, J. (1979).Structural studies of Art Styles. In Cordwell, J.M. (Ed.), The Visual Arts: Plastic and Graphic. 139-212. Mouton Publishers. Great Britain.
21. Frohardt, D. (1999). Teaching Art with Books Kids Love: Teaching Art Appreciation, Element of art, and Principles of Design with award-Winning Children's Books. Fulcrum Publishing. United States of America.
22. Bragg, L. A. (2003). Children's perspectives on mathematics and game playing. In L. Bragg, C. Campbell, G. Herbert & J. Mousley (Eds.), Mathematics education research: Innovation, networking, opportunity (Proceedings of the 26th annual conference of the Mathematics Education Research Group of Australasia, Geelong, (Vol. 1, pp. 160- 167). Sydney: MERGA.
23. Scoresby J., Shelton B. (2011). Visual perspectives within educational computer games: Effects on presence and flow within virtual immersive learning environments. Instructional Science, 39, 227-254.
24. Medina, E. (2005), Digital Games: A Motivational Perspective, Proceedings of DiGRA 2005.
25. Brown, E., & Cairns, P. (2004). A grounded investigation of game immersion. In CHI '04 Extended Abstracts on Human Factors in Computing Systems (pp. 1297-1300). New York, NY: ACM.

26. Bouvier, P., Lavoue, E., & Sehaba, K. (2014). Defining Engagement and Characterizing Engagement - Behaviors in Digital Gaming. Simulation & Gaming, 45(4–5), 491–507.
27. Rozendaal, M. C., Keyson, D. V., de Ridder, H., & Craig, P. O. (2009). Game feature and expertise effects on experienced richness, control and engagement in game play. AI and Society, 24(2), 123–133.
28. Barnett, L. A. (1990) Playfulness: Definition, Design, and Measurement. Play and Culture, 3, 319-336.
29. Lumby, J. (2011). Enjoyment and learning: Policy and secondary school learners' experiences in England. British Educational Research Journal, 37, 247-264.
30. Abdul, A. & Felicia, P. (2015). Gameplay Engagement and Learning in Game-Based Learning A Systematic Review. Review of Educational Research
31. Engestrom, Y. (1993). Developmental studies of work as a test bench ¨ of activity theory: The case of primary care medical practice. In S. Chaiklin & J. Lave (Eds.) Understanding practice: Perspectives on activity and context (pp. 64–103). Cambridge, MA: Cambridge University Press
32. Squire, K. D. (2002). Video Games and Education: Designing learning systems for an interactive age.

Functions and Uses of the Sarawak Malay Proverbs

Abang Patdeli bin Abang Muhi
Dewan Bahasa dan Pustaka, Malaysia

Abstract. Proverbs are the integral part of spiritual treasures of the culture and language of the people, the age-old wisdom and skills used by them. Proverbs are also an interesting pieces of popular wisdom and tradition belonging to any culture, which help us to foreground the values a shared beliefs held by a speech community. By applying Sperber and Wilson's Relevance Theory, this dissertation research aims to analyze the functions and uses of Sarawak Malays' proverbs which up to now received very little attention amongst the local researchers. Besides, the application of this theory has successfully explains how the proverbs allow the speaker to express his or her intentions, feelings and insights in an implicit way and tacit manner. The findings demonstrate that the main functions of the Sarawak Malays' proverbs are criticism, advice, warning, educating, sarcasm and insinuation which also reflect the wisdoms and wits of the Sarawak Malays. This research also offered an explanation of how the often ironical, metaphorical and figurative nature of the proverbs significantly affects their understanding. In addition, I have also studied the use of ellipsis in the familiar proverbs as well as found out the existence of various types of proverbs amongst the Sarawak Malays' community. Ultimately, the conclusions of the research are to attain a better comprehension of the functions and uses of the Sarawak Malays' proverbs

1 Introduction

Most probably, only a handful of us know about the existence of the Malay community in Sarawak. Furthermore, compared to other ethnic groups in the state, particularly the Ibans, the Sarawak Malay community apparently does not capture the interest of the local as well as foreign researchers, especially among the sociologists and anthropologists.

Since this paper deals with the Sarawak Malays' proverbs, it is relevant indeed to furnish a brief socio-cultural background of the community. The Malays are the second largest ethnic group in Sarawak after the Ibans, and they are different from the Malays in the Peninsular Malaysia or Sabah in most aspects except for their religion. The Sarawak Malays have their own local language and dialects. According to Abang Yusuf Puteh (1998), their cultures and outlook on life generally are almost totally different too from the Peninsula Malays.

Abang Yusuf Puteh (1998:8) quoted Alfred Russel Wallace's, observation on Sarawak Malays his visit to Santubong etc .in 1885: "In character the Malay is impassive. He exhibits a reserve, diffidence, and even bashfulness, which is in some degree attractive, and leads the observer to think that the ferocious and bloodthirsty character imputed to the race must be grossly exaggerated. He is not demonstrative. His feelings of surprise, admiration, or fear, are never openly manifested, and probably not strongly felt. He is slow and deliberate in speech, and circuitous in introducing the subject he has come expressly to discuss. These are the main features of his moral nature, and exhibit themselves in every action of his life...."

Distinctly language defines culture, and proverbs as an element of language essentially encapsulate the values, beliefs and wisdom of the community. The Sarawak Malay culture and language have their own proverbs which are yet to be properly collected, documented, analyzed and described.

This paper, arising from a larger research project, is the first attempt to collect, document, describe and analyze the Sarawak Malays' proverbs. The research has successfully collected, described as well as analyzed 1373 verses of proverbs.

In this paper I will focus on the definitions of the proverbs, the analysis of the proverbs by applying Sperber and Wilson's Relevance Theory, and the functions as well as uses of the proverbs among the Sarawak Malays community. I will also touch on who are the Sarawak Malays as a brief introduction to the community. The last part of my paper will include several conclusions and recommendations for further research on the Sarawak Malays' proverbs.

I earnestly hope that, by presenting this paper pertaining to the functions and uses of Sarawak Malays' proverbs, the community will be well-known, and will indeed excite the interest of the local and foreign researchers towards the community, especially in the domain of oral tradition.

2 Sarawak Malays Community in Brief

In order to study the traditional literature of the Malay of Sarawak, the oral tradition in particular, it is imperative to have a glimpse of the community's culture and language as both aspects are the main sources of my data on their proverbs.

In general, the Malays of Sarawak are different from the other Malays in other states in Malaysia, except for the fact that they are all Muslims. They speak their own language, which is Sarawak Malay dialect or locally known as *basa Melayu Sarawak*. In term of culture, they practice their own culture that in one way or another, do have some similarities with the Malays elsewhere in Malaysia.

Actually, *basa Melayu Sarawak* is the lingua franca for the multi-ethnic, multi-dialect and multi-cultural communities in Sarawak. According to Article 160 of the Federal Constitution, a Malay is defined as a person who professes the Muslim religion, habitually speaks Malay, and conforms to Malay customs. In the Sarawak Constitution, by contrast, the term Malay is less elaborate as they are merely on of the communities mentioned in a long list of natives originated from Sarawak, but, due to the fact that they profess the Muslim religion, habitually speak Malay, and conform to Malay customs, herewith the Malays of Sarawak are legally regarded as Malay as per the Federal Constitution.

In addition to the clause of Article 160 mentioned above, Abang Yusuf Puteh (1998:2-3) elaborates that:

"An examination of the term 'Malay' in Sarawak would invariably bring us to two interesting anthropological processes of conversion: firstly, from non-Malay to Malay, and secondly, from a Malay to Malay. In Sarawak, the phrase *masuk Melayu* (which ~~is~~ literally translates as to enter Malay), culturally and legally means a non-Malay to become and accepted as a Malay. It is a process for a non-Muslim to convert to Islam, usually upon marriage. The terminology *masuk Melayu* is of course a misnomer but which has been accepted throughout the ages. The children of a Malay from the *masuk Melayu* process are firstly Muslims and secondly registered as Malays. For all practical purposes the convert is a Malay whose rights include the eligibility to buy and own native lands."

Besides Islam, the main characteristic that distinctively differentiate them from the other groups which were considered as the Malay of Sarawak such as those who have converted to Muslim as well as the Arabs, Kedayan, Javanese, Bugis etc., is their language. Awang Hasmadi (1998) quoted Harrisson (1970:132) as saying:

> *"Istilah Melayu dan kaum Islam Sarawak sering bertukar ganti penggunaannya. Oleh yang demikian ada kalanya pengertian Melayu juga merujuk kepada penduduk Sarawak yang lain yang beragama Islam seperti kaum Melanau Islam."* (The usage of the term Malay and Sarawak Muslims are regularly inter-changeable. As such, sometimes the meaning of Malay is used to refer to those Sarawakians who are Muslims such as the Muslim Melanaus.) Hence, the term Malay used in this paper as defined above.

3 Definitions of Proverbs

One of the Sarawak Malays' oral tradition which authenticate the poetic, beauty and rhythmic nature of their language are the proverbs or *peribahasa*. According to Mieder (2004), generally, proverbs are known as short sayings of the folk which contain wisdom, truth, morals and traditional views in the form of metaphors, or fixed and memorable forms which are handed down from generation to generation.

Thomas Fuller as quoted by Richard Winstedt (1957) defined a proverb as "much matter decocted into a few words", meanwhile, Cervantes defined a proverb as "short sentences drawn from long experience".

For the purpose of this paper, and due to the fact that the Sarawak Malays share common cultural and religious values as well as language similarities with the Malays in Peninsular Malaysia, the interpretation of proverb as defined by Za'ba (1962:165) as follows will be the main reference:

> *Peribahasa itu ialah segala susunan cakap yang pendek yang telah melekat di mulut orang ramai semenjak beberapa lama oleh sebab sedap dan bijak perkataannya, luas dan benar tujuannya, dipakai akan dia jadi sebutan oleh orang sebagai bandingan, teladan dan pengajaran. Maka ialah juga yang dikatakan bidalan, pepatah, perbilangan dan perumpamaan kerana tiap-tiap satu ini cakap yang mengandungi segala sifat peribahasa yang telah disebutkan itu.*
> (A proverb refers to all sequences of short sayings commonly used among the people from years ago due to the beauty and wisdom of the words, the breadth and truth of its purpose, these sayings were used by the people as comparison, example and lesson. Thus, it is also known as aphorism, adage, saying and maxim as each of these sayings demonstrate every characteristic of the proverbs mentioned.)
> For the sake of retaining the accuracy of the meanings of the terms, in this paper the Malay terms of *pepatah* or *bidalan*, *perumpamaan*, and *simpulan bahasa* rather than the translated equivalent in English will be used. The term 'proverbs' or *peribahasa* therefore refers to all the generic terms of *pepatah* or *bidalan*, *perumpamaan, perbilangan*and *simpulan bahasa*.

4 Sources of Data

1. The main sources of the Sarawak Malays proverbs in this research are:
2. The Malays of Sarawak: A Socio-Political Portrait (Abang Yusuf Puteh, 1998)
3. Adat Perkahwinan Orang Melayu Sarawak Edisi Kedua (Abang Yusuf Puteh, 2008)
4. Several interviews with two reliable informants who are well-versed in the Malays of Sarawak culture, traditions and oral tradition, particularly *pantun*. One of the informants was an ex-teacher who has published three books on Sarawak Malays *pantun*. The other one was a Quran teacher as well as an *ex-bilal*.

5 Types of Sarawak Malay Proverbs

By applying the interpretation of Za'ba (1962) as mentioned earlier, this research discovered that the proverbs of the Sarawak Malay comprised of:

1. *Pepatah or bidalan* (A proverb which has a single meaning).
2. *Perumpamaan* (A proverb which has a double meaning).
3. *Simpulan Bahasa*(A phrase comprising of two or more words and established by usage and not immediately comprehensible from the words used).
4. Values of Beliefs.
5. Two-line pantoum.

The summary of the findings of this research is stated in

Table 1. The number of proverbs according to types

NO.	TYPES OF PROVERBS	TOTAL
1.	*Pepatah atau Bidalan*	266
2.	*Perumpamaan*	91

3.	*Simpulan Bahasa*	964
4.	*Nilai Kepercayaan*	46
5.	*Pantun Dua Kerat*	6
6.	*Perbilangan*	0
7.	*Lidah pendeta*	0
		1373

The findings of this research also showed that there are no proverbs that can be categorized as *perbilangan* or a proverb comprising of two lines or more whereby its content might become a rule, law or advice that might become a guide. This research also found that there are no proverbs that can be considered as *lidah pendeta*. Besides the findings stated in Table 1 above, this research also uncovered that they are six types of *simpulan bahasa* as shown in Table 2 below. The *simpulan bahasa* was categorized into six types based on word class used in the proverbs, namely noun, verb, adjective, proper noun, function word and preposition.

Table 2. The Number of Verses Based On Word Class

NO.	*SIMPULAN BAHASA* BASED ON WORD CLASS	TOTAL
1.	Noun	387
2.	Verb	324
3.	Adjective	223
4.	Proper Noun	18
5.	Function Word	5
6.	Preposition	7
		964

The findings of this research pertaining to Sarawak Malay proverbs, especially on the number and types of proverbs as well as the description of the meanings, is the first of its kind.

6 Analysis Based on Relevance Theory

The application of Sperber and Wilson's Relevance Theory (1995) for this research is due to the fact that proverbs played a pertinent role in the communication process, especially among the traditional Sarawak Malays community. Besides
that,proverbs are parts of the oral tradition that are commonly observed among the community.

According to Relevance Theory, communication is a process which involves two distinctive mechanisms of information processing devices, namely involving the modification of the physical environment of the hearer, and the creating or building of representations which are similar to the ones of the hearer.

Meaning to say, Sperber and Wilson (1995) stressed that, in an oral communication the hearer usually tries to modify the hearer's environment in order to entertain some certain thoughts that the speaker wants to share with the hearer. Hence, if that is the case, the second device will create or build several representations that are similar to the ones of the hearer. The two approaches of Sperber and Wilson as mentioned above are indeed suggesting two pragmatic questions. Firstly, the question of "what is communicated ?" and secondly "how the objective of the communication is achieved ?"

Pragmatically, this research applied Sperber and Wilson's Relevance Theory to analyze proverbs based on the justification that the meanings of the proverb fundamentally depends on the identification of the speaker's intention, and the inference of the implicit communication rather than decoding what is explicitly articulated. Nevertheless, this research also took in account the proposed definition of the term 'relevance' suggested by Sperber and Wilson (1995): "an assumption is relevant in a context if and only the assumption has certain contextual effect in that context". To this extent their definition of the term "relevance" is pragmatic to be applied for the analysis of Sarawak Malays proverbs as having certain contextual effect is the essential condition for the relevance of a proverb.

7 Functions of Proverbs Among the Sarawak Malays Community

Proverbs as part of the oral tradition of the Malays, including the Malays of Sarawak, play a significant role in the communication process among the community. Furthermore, proverbs are the product of language that contains two main aspects, the cognitive as well as emotive aspects.

Iman Budi Santosa (2016:viii) asserted the functions and uses of the proverbs as:

> *Peribahasa atau sering disebut aforisme, wujudnya adalah kelompok kata atau kalimat yang susunannya tetap untuk mengungkapkan pernyataan tertentu. Isinya adalah perbandingan, perumpamaan, nasihat, prinsip hidup, aturan tingkah laku, baik yang dipujikan atau dilarang dalam tradisi setempat. Konstruksi kalimat pun umumnya dibuat singkat, padat, puitis, agar memiliki daya tarik sehingga mudah dihapal dan diingat oleh masyarakatnya. Dalam sejarahnya, peribahasa yang berisi nasihat lahir dan digali dari simpul-simpul pandangan hidup yang bersumber pada agama, kepercayaan, mitos, religi, falsafah, serta ajaran para cerdik pandai/pujangga/wali/raja/datu di masa lalu yang terbukti ampuh menjadi pedoman hidup mereka.*

(Proverb or usually referred to as aphorism, its existence are in the form of words cluster or sentences which have a fixed sequence for articulating certain expressions. They contain comparisons, maxims, advice, principles of life, order of behavior, whether honored or prohibited within the local tradition.)

Generally, the construction of sentence is also short, concise, poetical, so that it possessed the power of attraction until it is easy to be memorized and remembered by its community. In its history, the proverb contained advice which originated and were unearthed from the inferences view of life rooted from the religion, beliefs, myth, philosophy, as well as the teachings of the intelligentsia / literati/saint/king/dignitary years ago that has been proven purposeful as their guidance of life.)

Pertaining to the functions and uses of the Malay proverbs of Sarawak, this research has focused on the analysis of the literary styles of the proverbs. The analysis proved that its literary style contains aesthetic and artistic values. Both values are capable in arousing numerous feelings of their readers and hearers, and simultaneously increase their knowledge as well as stimulating their power of thinking.

This research also found that the dominant literary style of the proverbs are metaphors as well as figurative, and which are mainly functioning as innuendo, sarcasm, insinuation and mockery in an indirect manner. The research also proved that, the proverbs functioned as a lesson to their readers and hearers.

As mentioned earlier, the functions of the proverbs are to impart lessons, advice, comments and criticism in a refined and well-mannered way. The analysis of the 1373 collected Sarawak Malay proverbs discovered the existence of several elements of lessons, advices, comments and criticisms such as follows:

1. Courteous (berbudi bahasa).
2. Cautious (berwaspada).
3. Cooperative (saling bekerjasama).
4. Good at keeping secret (tahan menyimpan rahsia).
5. Not to behave boastfully, greedily, enviously, maliciously, insolently and so on (tidak bersikap sombong atau meninggi diri, tidak tamak, tidak iri hati, dan tidak mengumpat, tidak menghina dan sebagainya).

The proverb *jaga mulut* (keep your mouth) which means to be guarded when talking, for instance, is a reminder for everyone to be courteous in order to ensure mutual relationship among the members of society, and simultaneously, to avoid offending or hurting their feelings.

Meanwhile, *buah masak belum tentu manis* (ripe fruit not necessarily sweet) which means those who are nice-looking are not necessarily having a good behavior; this reminds us to be constantly cautious, and not easily deceived by what we see, by their lovely appealing or their beautiful appearances. In short, not to judge a book by its cover.

They are proverbs that contained advice that remind us about the importance of mutual understanding, mutual cooperation and mutual responsibility among the community to ensure peace and harmony in the society. The proverb b*erat sama dipikol ringan samadijinjeng* (bear the same weight the same light) is one of the example.

This research also revealed the element of advice that requires us to be good at keeping secret, and refrain from disclosing it in order not to cause harm or virulence to others as well to assure that he or she is faithful and sincere to each other. One of such proverb is *sak pecah di perut jangan pecah di mulut* (let burst in the stomach do not burst in the mouth).

Besides the elements of advice and lesson of the proverbs as described earlier, this research also found out that they are proverbs that advised us not to disgrace our own family, and so on as such an act will only insult and humiliate ourselves. An example of such proverb is *jangan meludah ke langit kelak nimpak muka dirik empun* (don't spit into the sky will fall into his own face) which means do not disgrace your own parents or your family members as it will disgrace yourselves.

Other functions and uses of the proverbs discovered from this research are pertaining to the ethos and values which are related to their function as elements of didactics in line with Harun Mat Piah (1989) view that like *pantun* and *teromba*, the proverb also depicts the ethos and values of the community. These ethos and values normally become the norms of the community concerned, and not excluding the Malay of Sarawak community.

The analysis of the proverbs of the Malay of Sarawak, surfaced that the elements of ethos and values also existed as stated below:

1. Courteous and meritorious (berbudi dan berjasa).
2. Humble (*merendah diri*
3. Hardworking (*bekerja kuat*)
4. Thrifty (*berjimat cermat*)
5. Loyal (*taat setia*)
6. Respecting the laws (*Menghormati Undang-undang*)

The proverb *ada padi ada beras, ada budi ada balas* (there is no rice paddy cultivation there is no reply) for instance, reflects the Malays of Sarawak emphasis on the importance of gratitude and good deeds among its community members. Being grateful is part of the dominant and integral cultural values among the Malays of Sarawak community, and simultaneously it binds them to much larger values. Being grateful is also not merely giving and receiving in terms of material, however, it also pertinent to other matters pertaining to the ethic, moral, cultivated manners as well as customs of the community.

The Malays of Sarawak do not tolerate those who are arrogant or snobbish as well as boastful. *Kambin gambong* (snooty goat) is one of the examples of the proverb that describe a person who arrogant or snobbish, and proud or boastful. This also resembles the snooty and fierce appearance of the goat. Another proverb that describe a person who is boastful is *besar kelakar* (big words or brag), and the Malays dislike such behavior.

One of the proverbs that reflect the antipathy of the Sarawak Malays towards an act of extravagant is *tin-tin telinga tikus, sak gentin jangan putus* which literally means no matter how tight it is just don't let it break. This proverb reminds the members of the community to be thrifty in their spending, and to be frugal in everything they do.

The Sarawak Malay community also lays a great importance on the values of responsibility and accountability. Every member of the society is required to be responsible and accountable for whatever deeds they commit. Nevertheless, pertaining to the issue of reliability, particularly regarding honor and self-respect, the words or confession of a woman is more reliable. The proverb *banyak malu kepada perempuan sedikit malu kepada laki-laki* (a woman more reliable than man) describes that the evidence or the confession made by a woman regarding the wrong deeds of a man towards her such as molesting, illicit sexual intercourse or adultery, is more reliable. That proverb can also be interpreted, due to the fact that she is a woman she is the one who will bear the shame more than the man in certain cases especially pertaining to extramarital affairs.

This research also showed that there are certain proverbs which require every members of the community to be a responsible person. Those who commit wrongdoings or bad deeds should be held responsible for what they did. They should not *lepas tangan* (wash their hands clean) of what they did, and instead they should be *berani nebang berani mikol* (brave brave bear felling). In other words, he or she should be *berani buat berani tanggung* (dare to take bold responsibilities).

8 Uses of Sarawak Malays Proverbs

The uses of proverbs as mentioned earlier was defined as a way of saying something implicitly rather than explicitly supporting one's arguments using a popular wisdom. Moreover, proverbs are full of metaphors as well as considered as part of figurative language due to the fact they are an indirect way of conveying a set of assumptions pertaining to shared thoughts between the hearer and the speaker rather than stating or conveying them directly.

Proverb is also a unique feature of the Sarawak Malay language in which message, advice, lesson, comment, remark and criticism is indirectly communicated through figurative language and metaphoric phrases. Notwithstanding, the uses of proverb in communication may precipitate misinterpretation as well as misconception and confusion among the hearers if they are not familiar with the phrases as they contained implicit meanings that need to be interpreted wisely.

a. According to Za'ba (1962:165), they are uses of *peribahasa* or proverbs such as:
b. *Menghiasi karangan atau percakapan* (decorating essay or conversation).
c. *Menguatkan tujuan karangan atau percakapan* (enhancing the goals of essay or conversation)
d. *Dijadikan pedoman hidup kerana isinya yang besar dan luas* (used as a guide of life because the content is large and wide-reaching).

Based on Za'ba's definition on the uses of the Malay proverbs, this research revealed that, they are three main uses of the Sarawak Malays proverbs such as stated below:

8.1 Advice and Lessons

Most of the proverbs found in this research are contemplated as an advice to the hearers rather than aimed to criticize them. Such proverbs also served as a lesson to the hearers in order ensure that they will be more cautious in the future, and refrain themselves in repeating such mistake and so on. As an illustration of advice and lesson, we can see two examples of proverbs below:

a. *Besar pasak dari tiang* (unbalanced budgeting) which means more spending compared to income). This proverb served as an advice so that the hearer will manage his or her income wisely in order to avoid overspending

b. *Udah beruban masih berolah* (literally gray but still throwing a tantrum) is some sort of a lesson for an old man or woman not to misbehave as if he or she is still juvenile. In other words, the elders should show good examples to the young ones.

8.2 Reminders and Warnings

This research also revealed that, the proverbs also aim to give warning and reminder to the hearers in order to warn or prevent them from doing something which may cause problem or danger. The examples of such proverb are as follow:

a. Di*am-diam lepuk* (silent silent lionfish) which means enemy or someone who is taciturn should not be taken lightly. This proverb reminds and warns the hearer to be cautious always in order not to be deceived by the appearance of a taciturn person.

b. *Sik pisang bebuah duak kali* (once bitten twice shy) which means once being cheated, deceived or suffered a loss, he or she should be daunted or discouraged and learn non's lesson. This proverb is a reminder and warning to the hearer in order not to repeat the same mistake as well as not to be cheated or deceived twice for the same reasons and so on.

8.3 Criticism

The analysis of the Sarawak Malays proverbs disclosed that, most of the times, the speaker emits or utters a proverb with the aim of criticizing the hearer or accusing or ridiculing the hearer. The following proverbs are the illustration of this function.

a) *Berani kesak* (brave weaver ant) which means a person who is only brave or daring such as when fighting, when he or she is with a group of people. This proverb criticizes the hearer who is in fact a coward, and not daring to challenge others face to face, and when he or she is alone.
b) *Jaik perut (rotten stomach)* which suggests that the bad qualities of a person are not from the heart but from the bowels. According to Abang Yusuf Puteh (1998:178), one possibility of its origin may be the reference to faeces in the stomach. This proverb is criticizing the hearer who is having a rotten heart, and bad attention which are not healthy for the establishment of a relationship or for the intercommunication among the community.

9 Conclusion

As I have observed and analyzed, the existence of the long list of Sarawak Malays proverbs which I have collected, described and documented through this research, is a contribution to the Malay civilization, especially in the field of oral tradition.

This study showed that, the proverbs play an important role in the traditional communication process among the Sarawak Malays community which are not popularly known among previous researchers.

The analysis of the proverbs revealed that they described the good and bad behavior of the Malay community. In other words, the functions and uses of the proverbs in the Sarawak Malays community are to describe a man or woman's behavior and attitude such as terms of honor.

To conclude, this research, the first of its kind as far as the Sarawak Malays proverbs are concerned, is a stepping stone for further research in the field of oral tradition among the Sarawak Malays, in the domain of proverbs in particular.

References

1. Abang Yusof Puteh. *The Malay Mind*. Shobra Peblications Sdn. Bhd., Kuching, Sarawak.1997.
2. Abang Yusof Puteh.*The Malays of Sarawak:A Socio-Political Portrait*.Shobra Publications Sdn. Bhd., Kuching, Sarawak.1998.
3. Abang Yusuf Puteh. *A Profile of Sarawak Malays*. Penerbit Universiti Pendidikan Sultan Idris, Tanjong Malim. 2005.
4. Abang Yusuf Puteh. *Adat Perkahwinan Orang Melayu Sarawak Edisi Kedua*. Dewan Bahasa dan Pustaka, Kuala Lumpur. 2008.
5. Awang Hasmadi Awang Mois & Sanib Said (Penyunting). *Masyarakat Melayu Sarawak: Himpunan Kertas Seminat Budaya Melayu II*. Yayasan Budaya Melayu Sarawak, Kuching. 1998.
6. Harun Mat Piah. *Puisi Melayu Tradisional : Satu Pembicaraan Genre dan Fungsi*. Dewan Bahasa dan Pustaka, Kuala Lumpur. 1989.
7. Iman Budhi Santosa. *Peribahasa Nusantara, Mata Air Kearifan Bangsa*. DPP PDI Perjuangan, Jl. Diponegoro, Menteng, Jakarta. 2016.
8. R.O. Winstedt. *Malay Proverbs*. London, Murray.1967.
9. Sperber,D. & Wilson,D. Postface to the Second Edition of *Relevance:Communication and Cognition*. Oxford:Blackwell. 1995.

AUTHOR INDEX